Genetical Structure
of Populations

Genetical Structure
of Populations

KENNETH MATHER, F.R.S.
Honorary Professor of Genetics
in the University of Birmingham

LONDON

CHAPMAN AND HALL

First published 1973
by Chapman and Hall Ltd.,
11 New Fetter Lane, London EC4P 4EE

Printed in Great Britain by
Willmer Brothers Limited, Birkenhead

SBN 412 12140 9

© 1973 Kenneth Mather

Distributed in the U.S.A.
by Halsted Press, a Division
of John Wiley & Sons, Inc., New York

Library of Congress Catalog Card Number 73-13379

Contents

Preface   *page* vii

1. Variation and Selection   1

2. Types of Variation   4
   Heritable and non-heritable variation, 4
   Chromosomal variation, 6
   Genic variation, 14
   Variation in populations, 17
   Significance of the types of variation, 22

3. Equilibria   24
   The Hardy-Weinberg equilibrium, 24
   Inbreeding, 29
   Mutation, 37
   Selection, 44
   Competitive and frequency-dependent selection, 48
   Genetic load, 54

4. Theory of Variability   61
   Free and potential variability, 61
   Utilized and fixed variability, 65
   Two types of potential variability, 67
   Unequal gene frequencies, 75
   Dominance, 78
   Mutation and drift, 81
   Selection experiments, 84

5. Types of Selection   *page* 88
    Stabilizing selection, 88
    Directional selection, 92
    Disruptive selection, 96
    Types of environmental change, 97

6. Life-Cycles and Genetic Systems   101
    Modes of reproduction, 101
    Breeding systems, 106
    Segregation and recombination, 118
    Genetic systems, 125

7. Genic Balance and Genetic Architecture   129
    Heterosis and balance, 129
    Integration and inertia, 138
    Genetic architecture of characters, 143

8. The Consequences of Disruptive Selection   148
    Nature of disruptive selection, 148
    Polymorphism, 149
    Developmental plasticity, 157
    Divergence, 159
    Isolation, 163
    Isolating mechanisms, 169

9. Individuals and Populations   172
    Units of variation and selection, 172
    Human populations and societies, 178

References   182

Index   191

Preface

The genetical study of populations has many facets, ranging from the theoretical and often highly mathematical to that direct investigation of individuals in their wild habitats, which has come to be known as ecological genetics. All these approaches are concerned in one way or another with the interplay of heritable variation and natural selection and all are linked by the underlying genetical mechanisms upon whose properties this variation and its reaction to the various forms of selection must depend. One of the outstanding findings of genetics has been that these mechanisms, as Darlington showed when he first published his *Evolution of Genetic Systems* over 30 years ago, are themselves built up, adjusted and readjusted by natural selection. Because, therefore, of the control these mechanisms exercise over the manifestation, transmission and reassortment of variation the action of selection in the past must determine the capacity for response to selection in the future. Its past will thus mould the response of a population to any future force of selection that may come to be imposed on it, will frequently decide whether this response is a success or a failure in achieving adjustment to the new demands, and will affect the way in which successful adjustment is eventually achieved. And underlying the adjustment of the many characters, morphological, physiological and functional, that may be directly involved, is the further prospective readjustment of the genetic system itself.

In his pioneering work, Darlington was concerned especially with the chromosome system and the control of genetic recombination that depends on it. The same principle applies, however, to reproductive and breeding systems, to genic balance and to genetic architecture; and its full significance in all these respects emerges much more clearly when we turn to consider polygenic systems, with their capacity for carrying very large amounts of concealed variability and their unique property of giving smooth change of the phenotype and close adjustment under selection, as well as the ubiquitous occurrence of the continuous variation they characteristically mediate.

Though genetic systems and polygenic variation are by no means the whole of the picture, I hope that the approach I have adopted will serve to bring out their central place in our understanding of the genetical structure of populations and indeed of the interdependence of this structure and the biological properties of the organisms themselves. Although I hope that it may interest a wider circle, this book is intended primarily for more senior undergraduates and research students. I have, of course, assumed a basic knowledge of genetics and chromosome behaviour. I have deliberately restricted the mathematical content, but have included references to more mathematical treatments. Many of the results and phenomena I describe and discuss are capable of much lengthier documentation than it is reasonable to include, and in the often invidious task of choosing the illustrative references to give I have sometimes taken earlier papers in order to provide a measure of historical perspective, and sometimes more recent ones where these provide extensive references to earlier work; and in general I have, of course, tended to refer to the literature that is more familiar to me.

Certain of the figures are reproduced from earlier publications. I am indebted to the Royal Society, the Botanical Society of Edinburgh, the Society for Experimental Biology, Professor J. L. Jinks, Professor J. M. Thoday, and Messrs. Oliver and Boyd for their permission to do so in the various cases. Specific acknowledgement of its source is given below the figure in each of these cases.

February, 1973   K. M.

1 Variation and Selection

The individual members of a population of living organisms characteristically differ among themselves in their genetical constitutions. This has long been established from a variety of studies, covering both cytological observation and experimental breeding and made on a wide range of species, plant, animal and microbial (see for example Dobzhansky, 1951; Stebbins, 1950; White, 1954; Darlington, 1963), and the volume of evidence continues to grow apace. Our knowledge of the range of variation, even if not always of the precise nature of its genetical causation, is particularly detailed in the case of the species we know best – man himself (see Stern, 1960; Harris, 1971) but no species which has been examined adequately has failed to reveal genetical differences in its populations, and in some, notably *Drosophila* species, it is known to be very extensive.

Thus populations are not genetically homogeneous: on the contrary they are mixtures of genotypes. Nor are they constant in their genetical make-up. There has been, for example, a striking change over the past 20 years in the populations of certain bacteria which occur in our hospitals. Before the introduction of penicillin these bacteria were characteristically susceptible to the action of this antibiotic, but now a disturbingly high proportion of them are resistant to treatment with it, and they are resistant because of changes in their genetical constitution. The same is true of other bacteria and other antibiotics. It is also true of insects in relation to D.D.T. and we face agricultural as well as medical consequence of these changes. In the same way the introduction into our potatoes of genes for resistance to attack by the potato blight fungus (*Phytophthora infestans*) would appear to have been accompanied by a change in the genetical constitution of the fungal population. Other similar cases could be cited.

The importance of these examples is that in them we can see the cause of the genetical change. When a bacterial population is exposed to the action of an antibiotic, or a population of house flies to D.D.T., most will be killed; but any bacterium or fly which, because of its genetical constitution, is resistant to the treatment will survive to leave progeny sharing in greater or lesser degree the parental resistance, and the change will generally be cumulative over several generations of treatment. In the presence of antibiotic or insecticide the bacteria or flies differ in their capacities for surviving and contributing to posterity, that is in their Darwinian fitness, and the population changes in response to the selection which antibiotic or insecticide thus imposes. Being genetical in their causation, these changes are prospectively permanent: the populations have evolved to meet the demand of the new environment.

Since bacteria have very short life cycles, change is rapid under the impact of a powerful agency of selection such as an antibiotic. Even with insects the life cycle is sufficiently short for change to be relatively fast. With slower breeding species change will be correspondingly more difficult to discern; but comparative studies of populations living in diverse environments may still reveal differences attributable with some confidence to a recognizable agent of selection. Thus the incidence in negro populations of the gene which mediates the production of S-type human haemoglobin and which thereby confers on its possessor resistance to malignant tertian malaria, while at the same time producing sickle-cell anaemia in those homozygous for it, is relatable in a way we shall examine later to their risks of exposure to this type of malaria, see Allison (1955). We know too that plants of certain grass species growing on the spoil from mines show genetically determined resistance to metal toxity, whereas other individuals of the same species growing nearby but on unpolluted soil do not have this resistance, (see Bradshaw, 1971).

In all these cases the selective agent is clearly established. Other cases are known of genetical change where this is not so. Dobzhansky (1971) records the relative frequencies of certain types of chromosome in Californian populations of *Drosophila pseudoobscura*. These frequencies have changed markedly over the 30 years or so of his observations; but most of the changes have yet to be related to any corresponding alteration in the flies' environment. Nevertheless, few geneticists would doubt that these changes in the *Drosophila* population are the outcome of the action of selection, albeit selection stemming from agencies yet to be recognized; for the evidence both from direct observation in nature of the kind we have discussed and from the artificial imposition of selection in laboratory experiments, leaves no doubt that selection is much the most powerful agent of change in genetical constitution to which populations are exposed. If, therefore, we are to understand the genetical structure of populations we must not only examine the kinds of genetical differences that occur in them, the ways in which these differences arise, their interactions and their consequences for the individuals that carry them, but we must also bring selection explicitly into our examination as the most powerful agent in determining the fate of these differences and in moulding the structure of the populations. In doing so we must recognize that although we have so far seen selection as an agent of change, it can be an equally powerful opponent of change and, as we shall see, can even mould by its action the system of variation itself. The genetical study of populations, and with it the study of the mechanism of evolution, is chiefly the genetical study of variation and selection in their interplay with one another.

2   Types of Variation

Heritable and non-heritable variation

Differences in phenotype between individuals in a population may be traceable to differences in genotype or they may reflect non-heritable effects imposed on the individuals by differences in their environments. Our concern is with the former which are in principle transmissible to offspring, and not with the latter which will in general die away with the environments that brought them about. Environmental effects cannot, however, be entirely dismissed from consideration for two reasons. They may affect the way in which genetical differences display themselves; cases are known where a genetical difference fails to display itself at all unless the environment is of a particular kind and there are even more examples of a genetical difference displaying itself to a greater extent in one environment than in another. This means, of course, that the heritable variation detectable among the phenotypes of a population will in some measure depend on the environmental circumstances of that population, and in consequence the genetical response of a population to selection, which distinguishes directly between phenotypes and only indirectly between genotypes, will be to a corresponding extent conditioned by non-heritable effects. Furthermore, even where there is no specific dependence of a genetical difference on the environment in the display of its consequences, that is where the genetical and environmental differences are simply additive in the effects they produce on the characters, the greater the amount of non-heritable variation the greater the extent to which the phenotypic difference seized on by selection will be non-heritable and the less the efficacy of selection in affecting the genetical constitution of the population. Selection will, nevertheless, always produce its effects, albeit lesser effects the more the genetically determined differences are overlaid and obscured by non-heritable variation. Only where its genotype endows the organism with an adaptive plasticity which enables it to adjust itself precisely to the differing demands of diverse environments will the power of selection be circumvented without an increase in the variation displayed by the phenotype. In the extreme such adaptive plasticity could, of course, render all selection nugatory, but this would require a most precise system of compensatory adjustment in development, no approach to which has yet been detected in experiment.

The second reason why we cannot entirely ignore the effects of the environment is that these may stimulate heritable changes. The effects of ionizing radiations and many chemicals in stimulating a wide range of changes in the genetical materials are well known, though in general these agencies appear to be effective by raising the frequency of occurrence of the sort of undirected change, or mutation, that occurs even in the absence of specific exposure to their action. A few cases are, however, known where particular differences in the environment regularly produce specific changes which are then transmissable to offspring. Such behaviour is known in flax (Durrant, 1962), *Nicotiana rustica* (Hill, 1965) and possibly a few other plants; but it appears not to be widespread and its mechanism has yet to be elucidated, though it may depend on changes in ancillary materials of the chromosome.

With these reservations, however, we may turn our attention away from the effects of the environment and concentrate on changes and differences in the genetical materials themselves. Most heritable differences, are, of course, traceable to the nuclear genotype, though there is evidence from a range of species, mostly microbial or plant, of transmissable differences mediated by non-nuclear elements (Jinks, 1964). These differences have been observed to arise in the laboratory in a variety of ways and, as we shall see also to be the case with gene mutation, the changed forms in the plasmon are commonly disadvantageous, though some of them may at times confer an advantage on their possessors at least under the same conditions. Extra-chromosomal, or plasmatic, differences are, however, also known in nature having been found between different races of the willow-herb, *Epilobium hirsutum*, and between species of half a dozen other plant genera, including mosses and fungi as well as angiosperms. They have been observed between races of the mosquito, *Culex pipiens*, too. Such plasmatic differences indicate that, together with the nuclear genotype, the plasmon can play its part in the process of adaptation especially, it would appear, in the building up of barriers to crossing. These differences nevertheless appear to be exceptional rather than regular in nature and by comparison with nuclear variation of marginal significance to our understanding of variation in natural populations.

Chromosomal variation

Variation in the nuclear materials may be divided into two classes, chromosomal and genic (Fig. 1). Chromosomal variation involves whole chromosomes, or major portions of chromosomes, and is, in principle, detectable by its cytologically observable consequences. Genic variation, as the name implies, is variation at the level of the gene, and is generally incapable of detection by cytological observation. It is inferred from its effects on the phenotype. The two categories are not sharply distinct: obviously genic variation must depend on differences in the materials of the chromosome, and in some cases, such as Bar in *Drosophila*, a genic difference initially recognized by its effect on the phenotype, has been shown subsequently to be due to a small, but nonetheless cytologically detectable, alteration in the structure of a chromosome. Thus beyond a certain point the classification into chromosomal and genic variation becomes arbitrary. It is, nevertheless, a useful and convenient classification for reasons which will become apparent.

Taking chromosomal variation first, we can again recognize two basic types of variation, numerical and structural. Numerical variation, as its name implies, is variation in the number of chromosomes. Such changes may involve whole sets of chromosomes and

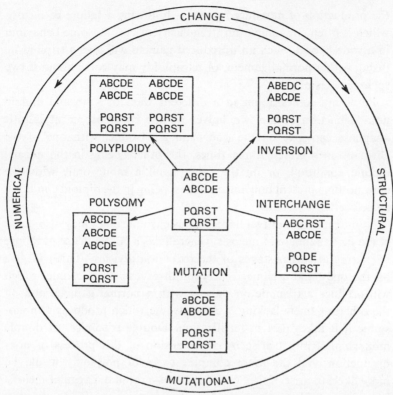

Fig. 1 Types of genetic change. Two chromosomes are shown, each initially in the diploid condition and each with five gene loci, A – E and P – T respectively. Tetraploidy is shown as an example of polyploidy, or balanced numerical change. Trisomy for chromosome A – E is shown as an example cf polysomy or unbalanced numerical change. Inversion is illustrated by reference to C – D – E and interchange by reference to D – E and R – S – T. Gene mutation is illustrated by change to a new allele a at locus A. (Reproduced by permission of the Botanical Society of Edinburgh and Messrs. Oliver and Boyd from K. Mather (1961), *Contemporary Botanical Thought*, 47 – 94, Oliver and Boyd, Edinburgh).

give rise to polyploidy, or it may involve only single chromosomes, or at any rate less than full sets, and give rise to aneuploidy or polysomy as it is sometimes called. Polyploidy can arise in a variety of ways, notably by somatic doubling of the chromosomes, which gives rise to tetraploidy when it occurs in a diploid organism, or by the production of unreduced gametes following a failure of meiosis which is often consequent on irregularity of chromosome behaviour in a wide hybrid. Such an unreduced gamete will give a triploid on fusing with a normal gamete, or tetraploidy may come about if two such gametes meet.

Polysomy comes about in a different way. In a normal diploid non-disjunction of the two halves of one chromosome at mitotic anaphase can give on the one hand a trisomic, with one of the chromosomes present three times, the others being in the normal diploid condition, or on the other hand a monosomic with one chromosome present only once, the rest being in the diploid condition as before.

Non-disjunction of two homologous chromosomes may also occur at the first anaphase of meiosis, generally as a consequence of failure of pairing at earlier stages of the reduction division. This will give on the one hand a gamete carrying an excess chromosome, which will produce a trisomic on pairing with a normal gamete, and on the other a gamete lacking a chromosome which produces a monosomic if it takes part in fertilization. Double trisomics, or double monosomics, might arise by an extension of this process of non-disjunction; but the more complex types of polysomy would be more likely to arise, if they occur at all, as a result of irregular meiosis in a wide hybrid or by a triploid producing gametes with an irregular number of chromosomes in excess of the normal haploid complement. Such irregular gametes appear, however, seldom to be viable in plants or to lead to viable zygotes in animals. Usually, indeed, the backcrossing of a triploid to a diploid gives nothing more than trisomic, or perhaps doubly trisomic, progeny.

Each set of chromosomes carries a full complement of genes. Thus polyploidy results in the addition of complete sets of genes and it is therefore not surprising that in plants, where polyploidy is common, polyploids in general show no marked departure from normal diploids in their somatic development and features. They are commonly distinguishable from the corresponding diploids, often by greater size, but the differences are neither profound nor marked. A hybrid polyploid, or amphidiploid as it is often called, will of course differ from its two ultimate parental diploids by combining characters from both or by displaying characters intermediate between them. Polyploidy is known in animals but it is much less common than in plants and our information about it is consequently less. It would seem that in man, at any rate, triploidy must have some disturbing effect on development since triploid foetuses become increasingly less common as gestation proceeds, (Polani, 1967; Carr, 1972). The cause of this disturbance, which could arise from some form of incompatibility between diploid mother and triploid foetus, rather than from upset intrinsic to the foetus itself, must remain a matter for speculation until further evidence is obtained.

Thus polyploidy, at any rate in plants where it is widespread, causes little upset of the somatic phenotype, and there is some indication that the additional robustness to which it can lead may endow its possessor with an extra ability to withstand colder and more rigorous environments. Its other effects may, at the same time, be profound. The odd-numbered polyploids, that is triploids (3x), pentaploids (5x), and so on, are incapable of undergoing regular meiosis and, in consequence, they are sterile or nearly so in normal sexual reproduction. Even-numbered polyploids, that is tetraploids (4x), hexaploids (6x), and so on, may behave with complete regularity at meiosis, though they do not always do so especially where they are derived from non-hybrid diploids. They may thus show some fall in fertility, but again this can be corrected by genetical adjustment as chromosome behaviour is in general under genic control (Rees, 1955) and in particular the regular meiotic behaviour of tetraploid and hexaploid wheat has been unambiguously referred to the action of a gene, or genes, in a specific chromosome arm (Riley and Chapman, 1958). The two chief effects of polyploidy are thus its ability to combine the full sets of two or more diverse parents in a fertile amphidiploid which breeds true for the complex hybrid condition; and to endow the hybrids, or indeed any other fertile polyploid, with the immediate capacity for maintaining themselves as lines distinct from their diploid forbears because of the sterility which characterizes the triploid progeny resulting from a backcross to their diploid ancestors.

B

By contrast with polyploidy, polysomy involves the addition – or loss – of individual chromosomes, rather than full sets. Since the genes carried by different chromosomes are not alike in function and balance, this means an inevitable imbalance of the genotype, some genes of which will be present in the nucleus in greater number than others. Polysomy thus produces disturbed development and if the imbalance is great, as where for example several chromosomes are present in the trisomic condition, or one is tetrasomic the rest being disomic, development is commonly so disturbed as to lead to a lethal condition. In plants the trisomic condition for a single chromosome is commonly not lethal, though it gives rise to a characteristic syndrome of distortion in development; indeed it was shown over 40 years ago that in *Datura stramonium* which has a haploid set of 12 chromosomes, the 12 single trisomics are individually and readily recognizable by the characteristic departures they show from the diploid phenotype. In *Drosophila*, trisomics for the small chromosome IV survive readily (as do monosomics for this chromosome) and trisomics for the X chromosome have been observed though they are of reduced viability (and of course grossly distorted sexual development); but trisomy for either of the large autosomal chromosomes II and III is lethal. In man also trisomics are known to survive to birth for only a few of the twenty-two autosomes and those that do survive are grossly abnormal, trisomy for chromosome 21 leading to mongolism.

Polysomy must thus be regarded as an unfitting, and commonly a grossly unfitting, aberration. Trisomics may nevertheless survive to breed, and indeed in some cases do, especially in plants. When they do so, the extra chromosome is passed on to a proportion of the progeny which thus inherit the trisomic condition and phenotype, but this proportion is commonly less than the theoretical one-half because of lagging and loss of the extra chromosome at first meiotic anaphase. The trisomic condition will thus tend to die out of a population because of its breeding behaviour as well as because of somatic unfitness.

Before leaving numerical variation, reference must be made to the special case of B chromosomes. These are chromosomes, usually

Table 1 B chromosomes in wild populations of plants (Darlington, 1963). The table shows the numbers of plants in the population found to carry the various numbers of B chromosomes. Data from two populations of *Tradescantia paludosa* are given.

Species		Number of B chromosomes								
		0	1	2	3	4	5	6	7	8
Poa alpina		—	—	3	2	11	4	9	1	4
Tradescantia	1.	36	2	7	3	—	—	—	—	—
paludosa	2.	45	15	20	11	1	—	—	—	—
Clarkia										
elegans		236	36	38	3	9	1	1	—	—

smaller ones, which are additions to the basic complement, and which are not essential to normal development, survival or breeding. They behave irregularly at meiosis and they vary, sometimes widely, in number in the species and populations in which they are found (Table 1). Though having no clear or specific effect on the phenotype, they may affect such general characters as vigour, fertility and the chromosome phenotype (see Müntzing, 1963; Jones and Rees, 1969; Ayonoadu and Rees, 1971) and Darlington (1963) regards them as a prospectively important source of variation especially in annual plants. It should be noted, however, that they are not known in the great majority of species, and even in species where they occur they are absent from many individuals and populations.

Structural variation of chromosomes involves no change in their number, but an alteration of the internal linear arrangement in one or more of them (Fig. 1). Chromosomes break and rejoin from time to time, sometimes spontaneously (which is another way of saying, for reasons which we do not know) and sometimes at increased frequency under the stimulus of ionising radiations or radio-mimetic chemicals. When broken they do not always rejoin, and if they do it is not always in such a way as to reconstitute the original arrangement. Many of the breaks and rejunctions produce chromosomes with either no centromere or more than one, and as such are incapable of passing through a series of mitoses. As a consequence they cannot survive and so need detain us no longer. Two basic types of structural change, each involving two breaks and two rejunctions, are however viable mitotically and require our attention. They may of course be combined with one another or with others like themselves to give more complex rearrangements.

The first of these basic types is interchange, in which pieces of two different chromosomes are exchanged; thus AB and CD exchange segments to give AD and BC. Interchange is also known variously as translocation and reciprocal or mutual translocation. The second basic type of structural change is inversion, in which a piece of a chromosome is inverted. Thus ABCD gives ACBD. The inverted segment may not include the centromere, or it may do so, in which case it can also be regarded as an interchange between the two arms of the one chromosome.

Neither interchange nor inversion characteristically affects the somatic phenotype of the individual carrying it, whether hetero-zygously or homozygously. Many cases are indeed known in *Drosophila melanogaster* where the phenotype is affected as a result of so-called position effects of the changes; but even in *Drosophila* many other structural changes show no such position effects and they are hardly known in other animals and plants. The characteristic effects of structural change are thus indirect and they stem from the consequences at meiosis of the rearrangements. When homozygous the re-arrangements do not affect the course of reduction division though they will obviously alter the linkage relations of the genes. When heterozygous, however, interchange gives rise to a ring (or chain) of four chromosomes at meiosis, with two of the four chromosomes of the association generally passing into each gamete. If these pairs are of chromosomes adjacent to one another in the meiotic association, the gametes will carry duplication for some of the genes and deficiencies for others, thus AB–BC–CD–DA give AB + BC in some gametes and CD + DA in others, or BC + CD and DA + AB. Such combinations are generally inviable, though cases are known where they survive to give abnormally developing off-spring as with those human mongols which, because of an inter-change, are effectively trisomic for chromosome 21 though having only 46 chromosomes in the complement (see Court Brown, 1967).

Gametes deriving alternate chromosomes from the meiotic association, of four, on the other hand, carry balanced sets of chromosomes, being either AB + CD or DC + DA, and these will normally be fully viable. Alternate chromosomes of the ring or chain must, however, have come from the same parental line and this means that these pairs of chromosomes can never effectively recombine in an interchange heterozygote. Effectively, therefore, interchange suppresses recombination between the different chromosomes involved in it, the two chromosomes becoming a single unit in inheritance. With more than a single interchange rings of six, eight and more can be built up, and groups of three, four or more chromosomes locked together in a single unit of inheritance in this way. In some species of *Oenothera* all fourteen of the chromosomes that the nucleus carries come together in a single ring at meiosis with the result that the genotype consists effectively of a single pair of allelic units, or complexes as they are called, in transmission from parent to offspring.

Inversions when heterozygous also suppress, or very substantially reduce, recombination of the genes that they contain. Single crossing-over between the two homologous but relatively inverted chromosome segments give two non-recombinant chromatids of the parental types which pass successfully into gametes and hence to offspring, but also two recombinant chromatids one of which has no centromere and the other two centromeres with the result that neither passes into a gamete. The recombinant chromosomes are thus lost and recombination is effectively suppressed: recombinants are in fact recovered only if two reciprocal (i.e. 2-strand) cross-overs occur within the inversion. This will happen only relatively rarely even in long inversions and virtually never in short ones.

Since with both inversion and interchange the effective reduction or suppression of recombination depends on the inviability of the chromatids or combinations of chromosomes that it produces, all structural heterozygotes prospectively show a loss of fertility. This may, however, be regulated and reduced by special mechanisms, which in the case of interchange, ensure that in the great majority of meioses disjunction is such as to bring alternate chromosomes of the ring together in individual gametes, and in the case of inversion lead to the egg at any rate always, or nearly always, carrying a non-recombinant chromatid, the inviable recombinant chromatids passing to one of the other products of meiosis where they can cause no harm. Thus while indirect, the consequences of structural variation may be marked and have an importance in populations to which we will return later.

Genic variation

Chromosome variants commonly reveal themselves in the mitotic chromosomes or, even more often, in the characteristic configurations to which they give rise at meiosis. Genic variants reveal themselves characteristically by their effects on the phenotype and bring about no disturbance at either mitosis or meiosis, except in the cases of genes which play a part in the actual control of nuclear division itself. Gene changes must, of course, involve the addition, loss or substitution of material, presumably nucleotides at the appropriate places in the DNA molecule, or the rearrangement of this material; but it is on too small a scale to be detectable by direct observation of the chromosomes or generally by disturbances of their behaviour at meiosis. Since meiosis is undisturbed, genic variants segregate in the typical Mendelian fashion, albeit the ratios that are recovered may be distorted by such factors as selective fertilization and differential viability. Thus gene changes are revealed, and must therefore be classified, by their action and effects on the phenotype.

One such important classification was introduced by Muller (1932) who based it on comparison of the action of the mutant gene with that of its normal or wild-type allele. Most mutant genes have basically the same action as their wild-type alleles but are less effective or less adequate in producing this action. These he termed hypomorphs and in the extreme, where the mutant is devoid of action, they become amorphs, which action-wise are the equivalent of deficiencies. A few genes while having basically the same action as their allele, produce a greater effect on the character, and these he called hypermorphs. A still smaller class have a totally different effect and are termed neomorphs. Thus most genic changes, at least of the type familiar from the classical Mendelian experiment, result in genes which merely do the same thing as their normal alleles, but do it less effectively if at all. From the point of view of the phenotypic economy they are degradation changes. Only a very few changes can be regarded as prospectively constructive in that they introduce a new action – new, at least, for genes at that locus – and it is of interest that in many, if not all, of those known in *Drosophila* the change would appear to involve a small duplication of the chromosome material at the locus in question. These findings are not surprising when it is recalled that a gene is a complex ordered arrangement of part of the DNA molecule. Change is much more likely to damage such an integrated structure so that it works less efficiently or not at all than it is to make it more efficient or to endow it with a new, effective capacity for action. It is, therefore, confusing and dangerous to take the classical type of mutation as a basis for discussion of the development of the new genes that must have come into being to control new functions during the course of evolution.

The effects of genic variation are displayed by all the characters, without exception, that an organism can show – anatomical, physiological, biochemical, capacity for resisting disease or predator, capacity of causing disease or securing prey, antigen and antibody production, chromosome behaviour, mating system and so on, including with particular reference to man intelligence and psychological characters. Furthermore, the effects of a gene variant on a character may be of any magnitude, ranging from those so drastic as to be incompatible with normal development and survival, to those so small as to be detectable only by the most refined experimental and statistical techniques. A single gene change may affect several or even many characters of the individual that carries it, though it would seem likely that in many such cases the syndrome of ultimate effects stems from a single change in the initial action of the gene. Thus, to give but one well-known example, a lethal gene in the rat affects the skeletal, dental, circulatory and respiratory systems, death ensuing from any one or more of these effects; but this pleiotropy of the gene, as it is called, stems from a single initial action in upsetting cartilage development during embryonic life (Grüneberg, 1938).

A complex of changes might, however, come about in another way. Operator genes, such as Jacob and Monod (1963) recognized in the bacterium *Escherichia coli*, exert their effects by controlling the action of a number of structural genes each of which is responsible for mediating the production of an enzyme. Change in an operator gene thus alters the action of a number of other genes simultaneously and in this sense produces a pleiotropic effect; though it should be borne in mind that first, the structural genes Jacob and Monod found to be controlled by a single operator were all concerned with a single biochemical pathway; and second, the occurrence of such a system in a single-celled organism is no guarantee that its exact counterpart will be found in higher forms with multicellular somata.

Just as a number of characters can be affected by the pleiotropic action of one gene, a single character may be affected, and affected in the same way, by a number of genes. Such polymeric genes, as they are called, were first demonstrated by Nilsson-Ehle (1912) in cultivated wheat and oats and the finding of these polymeric genes in sets of three suggests that their occurrence is related to the hexaploid nature of these species. Nilsson-Ehle's polymeric genes had major effects on the phenotypes of the wheat and oat plants and so were capable of being followed in inheritance by the Mendelian segregation to which they give rise. Systems of genes segregating together and affecting the same character cannot however always be followed in this way, where the effect of each individual gene's substitution is small by comparison with the total variation to be observed. Such systems of genes, each with an effect small in this sense, similar to the effects of its fellows and capable of supplementing them, characteristically mediate continuous variation in which the expression of the character ranges by imperceptible gradations between wide limits, with intermediate levels of expression the most common (see Fig. 6). Polygenic systems of this kind cannot be handled or analyzed by the classical Mendelian techniques except in special and unusual cases where very refined methods are available. Rather, biometrical methods must be employed making use of statistical measures, means, variances and covariances, in place of segregation ratios (Mather and Jinks, 1971). We shall consider continuous variation and polygenic systems in more detail in Chapter 4.

Continuous variation has been found in all characters of all organisms, where an adequate study has been made. Since it affects the degree of expression of the character in a quantitative way it is often referred to as quantitative variation, and the characters as quantitative characters. This latter is however a misnomer since the same character can, even in a single set of observations, show continuous variation at the same time as the discontinuities in expression typical of the segregation of the classical gene pairs of major effect. Each of the groups into which the individuals are divided by the major genic discontinuities then shows continuous variation among its members. Continuous variation is as much a feature of wild as of laboratory populations and groups.

Variation in populations

To make a complete survey of the occurrence of all types of variation in even a small sample of individuals from a wild population would be so formidable a task as to be virtually impossible, unless the range of characters, in respect of which genic variation was to be surveyed, was very sharply restricted. There is every reason to believe, however, that all the types of variation that we have discussed occur, and indeed occur characteristically, in populations of all species. A few examples, some of them longstanding in the literature, will serve to make this point. As we have already had occasion to note, polyploidy is a well known and common phenomenon within species of plants (see for example Darlington, 1963; Stebbins, 1950). It is less common in animals but has been reported in a number of species (Lewis and John, 1963), including of course *Drosophila*. Data for the frequencies of chromosome variants are given by Fankhauser (1941) for an experimental population of newts, as set out in Table 2. Carr (1972) reports that about one-fifth of aborted human foetuses are polyploid, most being triploid but some tetra-

Table 2   Chromosome variants in tomatoes (Rick, 1945) and newts (Fankhauser, 1941), from Darlington and Mather (1949).

Type of Variant	Tomatoes (out of 66 unfruitful plants from 55 000 examined)	Newts (out of 1074 examined)
Haploid	2	1
Triploid	45	10
other Polyploids	3	6
Trisomics	2	1
Total	52	18

ploid also. Though most are aborted, occasionally a triploid foetus is born alive. Carr also summarizes data for other numerical abnormalities in human abortions and concludes that at least 90% of the foetuses which carry them will be lost spontaneously before twenty weeks of gestation. The different prospects of survival are revealed too by studies such as that of Polani (1967) who reports numerical variation in the general population but shows that it is commoner among stillbirths, so demonstrating yet again that numerical variation has such grossly upsetting effects as often to be incompatible with survival of the foetus to full term (Table 3).

Table 3   Frequency of Chromosome abnormalities in man (Polani, 1967).

Frequency in Percent	Trisomics	Sex Chromosome Abnormalities	Triploids	Others	All
At Conception (estimated)	1·37	0·78	0·50	0·64	3·29
In Survivors at Birth	0·19	0·18	0·00	0·04	0·41
Prenatal as Abortion	86%	77%	100%	94%	88%

A further study of numerical variation is of interest as it relates to a specific character. Rick (1945) found 66 unfruitful individuals among a group of 55,000 tomato plants. He sought to determine the cause of sterility in each case with the results shown in Table 2, which brings out well the variety of causes that can lead to a common effect.

Structurally changed chromosomes are not always detectable by inspection of the mitotic chromosomes: it is often necessary to observe the special meiotic configurations they produce when heterozygous to be sure of their occurrence. In *Drosophila* and related genera, the polytene chromosomes of the salivary glands facilitate the detection especially of inversions, chiefly through the homoglous pairing configurations that they show, but also because of the characteristic banding patterns whose rearrangements can be detected even when homozygous. Inversion variation has proved to be widespread in populations of *Drosophila* species and related forms. The two closely related species *Drosophila psuedoobscura* and *persimilis* between them show twenty-two different arrangements in their chromosome III of which several occur in each population, the precise sample varying from one population to another (Dobzhansky, 1951). This is one of the more extreme cases of inversion variation in these flies but inversions commonly occur in the chromosomes of other species as well and are very frequent in some of them.

Interchange variation is known to be widespread in man. We have already had occasion to refer to the interchange which involves chromosome 21 and leads to mongolism through a trisomy concealed under the normal number of 46 chromosomes and Jacobs, Frackiewicz and Law (1972) record a frequency of 1·5 per 1000 for all recognizable rearrangements taken together among new born babies. They further found a mutation rate of at least 1 to 2 per 10 000 for these rearrangements in live human births. This figure is for various reasons almost certain to be an underestimate. Turning to other species interchange is well known in populations of grasshoppers for example, (White, 1954; Lewis and John, 1963), and is present in the plant *Campanula persicifolia* on a wide scale (Darlington, 1937) while in many species of *Oenothera* and in some other species such as *Rheoe discolor* its presence is the basis of the permanent hybridity which characterizes these forms (see Chapter 6). Variation in respect of B chromosomes has already been described, and some illustrative samples taken from Darlington (1963) given in Table 1.

When we turn to genic variation, we can recognize that it falls into several different categories: some genic variation, such as that mediating the common blood group systems in man (ABO, MNS, Rh. etc.), is present in all populations, which have two or more genotypes all reasonably common and which are consequently described as being polymorphic for these characters (Ford 1945, 1971). Man is polymorphic for other characters too, such as the ability to taste phenylthiocarbamide, and for certain biochemical features. Indeed, electrophoretic studies of arbitrarily selected proteins, including enzymes, have revealed that man is polymorphic for the genes at over one-quarter of some 50 or so loci that have proved to be involved in the control of these proteins (Harris, 1971). This type of polymorphism has also been found in other animal species, and the same high proportion of polymorphic loci has been observed in forms of *Drosophila*, mice and the horseshoe crab, *Limulus* (Selander *et al*, 1970). Other polymorphisms are known, sometimes associated with specific situations like industrial melanism and mimicry, in various insects and snails (see for example Kettlewell, 1956; Clarke and Sheppard, 1960, 1962; and various chapters in Creed, 1971). Perhaps the most common polymorphisms however, are those associated with the control of breeding, whether by separation of the sexes as in many animals or by the variety of incompatibility systems in plants which we will have occasion to discuss in later chapters. Polymorphisms of one kind or another are indeed widespread in both plant and animal species.

The second type of major genic variation is the occasional occurrence of abnormal individuals as a result of the presence in the population of deleterious mutant genes. These are well known in man where indeed they are the cause of upsets like haemophilia, phenylketonuria (with its associated mental deficiency), Friedreich's ataxia, albinism, and so on. Each condition is a relatively rare occurrence, phenylketonuria occurring about once in every 40 000 births and haemophilia once in every 12 000 males, to mention but two of the conditions. Yet, taking into account all such genically determined abnormalities, they must amount to a far from negligible fraction of the population. It is difficult to estimate this fraction reliably because of our insufficient knowledge of the part genes play in the determination of human abnormality; but Stevenson (1961) has estimated that 4 % of births in the population of Northern Ireland show anatomical, physiological or psychological abnormalities, and that at least a quarter of these do so for genetical reasons. In other words, at least 1 % of births, and possibly up to four times that figure in this population are abnormal from genetical causes. Taking into account conditions that manifest later in life, Berry (1972) gives the incidence in a population as up to 6 %. Since most of the genes responsible for these upsets are recessive and, as we shall see in the next chapter, heterozygous carriers of normal or near normal phenotype are in such cases much commoner than the abnormal homozygote, heterozygosity for these deleterious genes must be very common. Indeed Muller (1950) has calculated that each of us must be heterozygous for several of them.

Recessive lethal genes are commonly found in heterozygous conditions in populations of *Drosophila* species and have escaped more widespread detection in other species presumably because of lack of the special stocks needed to test for them. Not all genes that affect viability are, however, completely lethal. Of some 3000 chromosomes tested in *Drosophila willistoni* (Dobzhansky, 1951) over 35 % carried genes that were lethal or semi-lethal. About half of the remainder carried other genes affecting the viability of flies homozygous for them. A few actually raised viability by comparison with that of siblings heterozygous for a common tester chromosome; but most reduced it to a greater or lesser degree. Genes upsetting characters other than viability and thereby producing detectable abnormalities were also observed in the experiments on *D. willistoni* and are indeed well known in populations of *Drosophila* species. They have been found in a wide range of other animals and plants too. They are picked up from the population almost always in heterozygous condition, the homozygotes seldom being found, partly no doubt because of their intrinsic rarity of occurrence but also because, being abnormal, homozygous individuals must generally fail to last for long in the competitive struggle for survival.

The final type of genic variation in populations is that of polygenic systems mediating continuous variation. As we have already seen, continuous variation is ubiquitous; and while in all cases a proportion, and in some cases the major part, of such variation is non-heritable, the evidence indicates that polygenic variation must be equally ubiquitous. It has been demonstrated in many populations of *Drosophila* species taken from the wild and has been studied intensively in the poppy *Papaver dubium* (see Lawrence, 1972; Gale and Arthur, 1972). It has been found in many other species of flowering plants (see Mather 1953), in fungi (Pateman and Lee, 1960; Croft and Simchen, 1965; Simchen, 1966), in many mammals including monkeys (see Mather 1953); indeed it has appeared wherever an adequate search has been made, and its properties are clearly of prime importance to our understanding of the genetical structure of populations. It is well known from many investigations since Galton's time to be a commonplace in man.

Significance of the types of variation

All these types of variation, chromosomal and genic, must be taken into account in considering populations, but their consequences are so different that they must vary in the roles they play in a population and especially in the significance they have for the understanding of the mechanics of evolution. Polyploidy, especially following wide hybridization, has obviously been of considerable importance in plants but it can of course have played no part in the evolution of the many genera whose species all have the same number of chromosomes. This serves to emphasize that while polyploidy provides the means of combining different genomes in a single species which thereafter breeds true for the combination, it still leaves open the basic question of how these genomes evolved in the first place.

In addition to polyploidy the comparative study of the chromosomes in many species and particularly the course of meiosis in species hybrids shows that related species commonly differ structurally in their chromosomes. Yet the absence of any direct effect of structural change on the phenotype indicates that these structural differences between species are secondary rather than primary to their evolutionary separation.

Polysomy usually has such upsetting effects on development that it would hardly seem likely to have been a major source of the variation from which adaptations and evolutionary changes are built up. Similarly, the lethal and other deleterious gene changes that we find in populations seem hardly likely to be a raw material of evolution, except perhaps where special mechanisms such as that found in *Oenothera* species depend for their working on the failure of certain genetic combinations (see p. 120).

These types of variation must be regarded as evidence that mistakes do occur, that the chromosome mechanism occasionally fails to achieve perfect transmission of the chromosomes, yielding for example, trisomics and monosomics by failure of disjunction during mitosis and meiosis, and that the mechanisms of genic replication does not always work perfectly to give exact replicas of the integrated gene structures. Despite these indications that some variation is little more than genetical detritus, the variation that we see must include prospectively valuable types and in particular there can be no doubt that genic variation is the basis of adaptive and evolutionary change, for no other variation is so widespread in populations or so characteristic of the differences between species. Furthermore, when species are crossed and their hybrids are successfully bred to give F_2s, it is clear that the gene differences are many and characteristically of the kind that give rise to continuous variation (Mather 1943, 1953). Nor is this surprising as polygenic systems, with their capacity for building up big differences by the accumulation of many genes of individual small but cumulative effects, clearly offer the means of securing that fine adjustment essential to the successful adaptation which Darwin recognized as the basis of evolution. Our consideration of the genetic structure of populations must, therefore, centre on genic variation and must pay particular attention to the special properties of polygenic systems.

3 Equilibria

The Hardy-Weinberg equilibrium

In respect of two allelic genes, G and g, each individual in a population falls into one or other of three genotypic classes, GG, Gg and gg. Let the frequencies of these classes in the population be respectively p, q and r where $p+q+r = 1$. Then $p+\frac{1}{2}q$ of the genes will be G and $r+\frac{1}{2}q$ will be g. We may denote these gene frequencies by $u = p+\frac{1}{2}q$ and $v = r+\frac{1}{2}q$ where of course $u+v = 1$.

Now assume that the contribution that each individual makes to the next generation is independent of its genotype in respect of $G - g$; and assume further that the individuals come together at random in mating. Then the frequencies of the different types of mating will be as shown in the second column of Table 4, the third,

Table 4 The Hardy-Weinberg Equilibrium.

The frequencies of the various types of mating and their contributions to the GG, Gg and gg classes among the offspring, in a parental population consisting of p GG; q Gg; r gg.

Type of Mating	Frequency	Contribution to		
		GG	Gg	gg
GG × GG	p^2	p^2	—	—
GG × Gg	$2pq$	pq	pq	—
GG × gg	$2pr$	—	$2pr$	—
Gg × Gg	q^2	$\frac{1}{4}q^2$	$\frac{1}{2}q^2$	$\frac{1}{4}q^2$
Gg × gg	$2qr$	—	qr	qr
gg × gg	r^2	—	—	r^2
All Offspring		$(p+\frac{1}{2}q)^2$ $= u^2$	$2(p+\frac{1}{2}q)(r+\frac{1}{2}q)$ $= 2uv$	$(r+\frac{1}{2}q)^2$ $= v^2$

fourth and fifth columns of which show the contributions that each mating type will make to the three different genotypes in the next generation. Thus to take an example, since a proportion q of the population is Gg and a proportion r is gg, the mating Gg × gg will occur in $2qr$ of cases, and since it gives rise to equal numbers of Gg and gg offspring, its contribution to each of these classes in the next generation will be qr, as shown in line five of the Table.

The frequencies of GG, Gg and gg in the next generation will be the sums of columns three, four and five as shown at the foot of the Table, and it will be seen that they reduce to GG u^2; Gg $2uv$; gg v^2. We could in fact have derived this result more quickly by recognizing that our assumption of random mating among the parents implied random assortment of the gametes in fertilization. Then as u of the gametes carry G and v carry g, G will meet G to give GG in $u \times u = u^2$ of cases, g will meet g to give gg in v^2 of cases and G will meet g to give Gg in $2uv$ of cases. It then obviously follows that as the gene frequencies in the offspring population are unchanged, being

$$u^2 + \tfrac{1}{2}(2uv) = u \quad \text{and} \quad v^2 + \tfrac{1}{2}(2uv) = v$$

respectively, the third generation derived by random mating among the offspring population, will once again carry the genotypes with the frequencies GG u^2; Gg $2uv$ and gg v^2. The population will continue with this constitution, generation after generation, so long as the assumptions of random mating and independence of the contribution to the next generation of the genotype in respect of G, g remain valid. The population is thus in equilibrium. This result was derived some sixty years ago almost simultaneously by Hardy in Cambridge and Weinberg in Germany and is consequently known as the Hardy-Weinberg equilibrium.

It will be observed further that this equilibrium is achieved after a single generation of random mating, for given that u of the genes were G and v were g, it makes no difference how these genes are distributed between homozygotes and heterozygotes in the parents. To take an arithmetic example, let $u = v = \dfrac{1}{2}$. Then the initial population could have consisted of GG $\dfrac{1}{4}$; Gg $\dfrac{1}{2}$; gg $\dfrac{1}{4}$ or, if it had

c

included only homozygotes, of GG $\frac{1}{2}$; Gg 0; gg $\frac{1}{2}$ or of an infinite

variety of intermediate constitutions such as GG $\frac{1}{3}$; Gg $\frac{1}{3}$; gg $\frac{1}{3}$ or

GG $\frac{2}{5}$; Gg $\frac{1}{5}$; gg $\frac{2}{5}$. No matter what its constitution had in fact been,

however, the offspring would be GG $\frac{1}{4}$; Gg $\frac{1}{2}$; gg $\frac{1}{4}$ and these would

be the equilibrium frequencies of the genotypes.

The distribution of the M – N blood groups in three samples of people from the United States, white, negroes and amerinds, as recorded by Landsteiner and Levine (see Mather, 1951), are shown in Table 5. Use of the two anti-sera, anti-M and anti-N, allows the three genotypes to be distinguished; for red cells from MM react only with anti-M, and NN only with anti-N whereas those from the heterozygote MN react with both. The frequency of gene M is easily found as $u = p + \frac{1}{2}q$. In the case of the sample of whites this gives

$$u = \frac{139 + \frac{1}{2}(285)}{532} = 0.529$$

The Hardy-Weinberg expectations for the three genotype frequencies are then easily found and can be compared statistically with the frequencies observed. In none of the three cases, white, negro or

Table 5   The M–N blood groups in three samples from the United States (Landsteiner and Levine, quoted by Mather, 1951).

Numbers of individuals falling into the three classes distinguished by the alleles M and N.

Sample		MM	MN	NN	Total	*u*
		\multicolumn{3}{c}{Genotype}		
White	O	139	285	108	532	0·529
	E	148·9	265·1	118·0		
Negro	O	50	86	45	181	0·514
	E	47·8	90·4	42·8		
Amerind	O	123	72	10	205	0·776
	E	123·3	71·4	10·3		

O = Observed   E = Expected at the Hardy-Weinberg equilibrium.

Table 6 The Hardy-Weinberg equilibrium for sex-linked genes.
The female is assumed to be the homogametic sex.

Frequencies	Females			Males	
	GG	Gg	gg	G	g
Parents	p	q	r	s	t
Offspring	$s(p+\frac{1}{2}q)$	$s(r+\frac{1}{2}q)$ $+$ $t(p+\frac{1}{2}q)$	$t(r+\frac{1}{2}q)$	$p+\frac{1}{2}q$	$r+\frac{1}{2}q$
Equilibrium	$\frac{1}{2}u^2$	$\frac{1}{2}(2uv)$	$\frac{1}{2}v^2$	$\frac{1}{2}u$	$\frac{1}{2}t$

At equilibrium $u = p+\frac{1}{2}q+s$ $\qquad v = r+\frac{1}{2}q+t$

amerind is there a statistical departure from expectation. So, although the amerinds differ from the whites and negroes in their frequencies of genes M and N, all the groups accord satisfactorily with the Hardy-Weinberg expectation, and there are therefore no grounds for postulating any departures from random mating or differences in fitness among the genotypes.

It is not difficult to extend the Hardy-Weinberg result to multiple alleles. With two alleles, the zygotic frequencies $u^2+2uv+v^2$ can be obtained by expanding the binomial $(u+v)^2$. With n alleles, having gene frequencies $u_1, u_2 \ldots u_n$ and $S(u) = 1$, the frequencies of the zygotic genotypes are given by $(u_1+u_2+\ldots+u_n)^2$.

As originally derived the Hardy-Weinberg equilibrium applies to autosomal genes. Where a gene is sex-linked the position is different in that only the homogametic sex can show the three genotypes GG, Gg and gg, the heterogametic sex having only two, G and g. Let the frequencies of these five genotypes in the population be p, q, r, s and t respectively, and for the sake of convenience in exposition let us take the heterogametic sex to be male. (Table 6). Now males derive their sex-linked genes solely from their mothers. Thus all sons of GG females will be G as will half the sons of Gg. Then in the next generation s' (the prime distinguishing the offspring generation from the parental) will be $p+\frac{1}{2}q$ and similarly $t' = r+\frac{1}{2}q$. Now at equilibrium the proportion of males carrying G must be constant over the generations and $s = s' = p+\frac{1}{2}q$. Then at equilibrium $s = p+\frac{1}{2}q$

and $t = r + \frac{1}{2}q$, or in other words the gene frequencies are the same in two sexes. Furthermore, when $s = p + \frac{1}{2}q$, it will be seen from the table that $p = s(p + \frac{1}{2}q) = s^2$ while $r = t(r + \frac{1}{2}q) = t^2$ and $q = s(r + \frac{1}{2}q) + t(p + \frac{1}{2}q) = 2st$. Thus setting the joint frequency of G as u and of g as $v = 1 - u$, the equilibrium frequencies are as shown in the bottom line of Table 6, the coefficient of $\frac{1}{2}$ being introduced on the assumption that the sexes are equally frequent, in order that the frequencies of the five genotypes sum to unity. The frequency of G taking the sexes together is $u = p + \frac{1}{2}q + s$. At equilibrium this is of course $\frac{1}{2}(u^2 + uv) + \frac{1}{2}u$ which is twice the frequency of G males. Since $s' = p + \frac{1}{2}q$ it is clear that in contrast to autosomal genes, equilibrium is not attained for sex linked genes after a single generation of random mating. It is, however, approached fairly rapidly and five or six generations of random mating will give a good approximation to it.

The sex-linked genes for yellow versus black or grey coat colour in the cat will serve to illustrate the equilibrium distribution. All the genotypes are distinguishable by direct inspection of the phenotype, the heterozygous female being tortoise-shell in colour.

A sample of stray cats from London has been classified by Searle (Falconer, 1960) with the results set out in Table 7. There are $(2 \times 7) + 54 = 68$ yellow genes out of the $2 \times 338 = 676$ genes at this locus in the females. Thus v, the frequency of the yellow gene, is $68/676 = 0 \cdot 1006$ in the females. In the males v is $42/353 = 0 \cdot 1190$, or if we take the two sexes together it is $(68 + 42)/(676 + 353) = 0 \cdot 1069$. Using this joint estimate of v, which of course also gives $u = 1 - v = 0 \cdot 8931$ as the frequency of the $+$ allele, the Hardy-Weinberg expectations are easily found to be as shown in the lower line of the Table. The numbers observed are in satisfactory agreement statistically with these expectations, thus showing that there are no grounds for postulating departure from a Hardy-Weinberg equilibrium in this group of cats. The heterozygous phenotype, tortoise-shell, can obviously occur only in females. At the same time yellow is commoner in males, among whom it occurs with frequency u by contrast with the frequency among females of u^2. In man, colour-blindness due to a recessive sex-linked gene, is similarly commoner in males than

Table 7 Frequency of the sex-linked gene for yellow colour in a London population of cats. (Searle's data quoted by Falconer, 1960).

Number of cats	Females				Males		
	++	+y	yy	Total	+	y	Total
Observed	277	54	7	338	311	42	353
Expected ($v = 0{\cdot}1069$)	269·60	64·54	3·86	338·0	315·26	37·74	353·0

Frequency of the gene y
In females $v = [(7 \times 2) + 54]/(2 \times 338) = 0{\cdot}1006$
In males $v =$ $42/353$ $= 0{\cdot}1190$
Overall $v = [(7 \times 2) + 54 + 42]/[(2 \times 338) + 353] = 0{\cdot}1069$

females. Pickford has recorded a frequency of 7·8 % in men and one of 0·65 % among women (Darlington and Mather, 1949), the female figure approximating to the square of the male frequency, as expected. The heterozygous women are of normal phenotype, being distinguished from their homozygous normal sisters only by their capacity for producing as many colour blind as normal sons. Taking the male and female frequencies of colour blindness together we can arrive at a joint estimate of

$$v = \tfrac{1}{2}(0{\cdot}078 + \sqrt{0{\cdot}0065}) = 0{\cdot}081$$

for the frequency of the gene for colour-blindness and go on to calculate that a further

$$2uv = 2 \times 0{\cdot}919 \times 0{\cdot}081 = 0{\cdot}15 \text{ or } 15\%$$

of females should be heterozygotes, the remainder being homozygous for the normal allele.

Inbreeding

In considering the distribution of both autosomal and sex-linked genes in populations we have explicitly assumed both random mating and the absence of differences in fitness among the genotypes. We have implicitly assumed, too, that the genes themselves do not change, G always remaining as G and g as g as they are transmitted from generation to generation. We must now examine the effects of departure from random mating, of change in the genes, that is mutation, and of differences in fitness leading to selective differentials among the genotypes.

Taking the mating system first, inbreeding, that is the preferential mating of relatives, leads to homozygosis, other things being equal. The simplest and most rapid form of inbreeding open to diploid organisms is the regular self-fertilization which is found in many species of plants, both cultivated and wild. Homozygotes breed true when selfed, and individuals heterozygous for a pair of alleles give half their progeny homozygous for one or other allele, only the other half remaining heterozygous, as indeed Mendel showed. Thus the proportion of heterozygotes is halved in every generation, the proportion of homozygotes rising accordingly. As an example, consider a population starting as an F_2. Its proportion of heterozygotes is $\frac{1}{2}$, which falls to $\frac{1}{4}$ in F_3, to $\frac{1}{8}$ in F_4, and so on. The proportion of homozygotes rises in the series $\frac{1}{2}, \frac{3}{4}, \frac{7}{8}$, etc. These homozygotes are of course of two kinds according to which allele they carry, but both kinds share the property intrinsic in homozygosity of being true breeding if the self-fertilization continues. Clearly populations under a regime of selfing must rapidly come to consist of homozygotes, though these homozygotes may differ genetically one from another. If, therefore, we raise progenies from a number of individuals taken from a population of a self-pollinating species of plant, we may expect to find heritable as well as environmentally determined differences between the progenies, but only non-heritable variation within them. As a consequence, selection among the progenies will be effective, but not selection within them: and this is indeed what Johannsen (1909) found in his classical experiments with pure lines of french beans, and which others have later observed in wild plants (for example, Haskell (1949) with *Stellaria*).

Self-fertilization is not possible for most animal species, where the sexes are separate. Brother-sister mating, or sib-mating as it is perhaps more conveniently called, is thus as close a form of inbreeding as is possible. Like selfing, the regular mating of sibs leads to homozygosis though distinctly more slowly, the proportion of homozygotes in a population starting as an F_2 rising according to the series $\frac{1}{2}, \frac{2}{4}, \frac{5}{8}, \frac{11}{16}, \frac{24}{32}$, etc. Regular mating of offspring with parent leads to homozygosis as rapidly as sib-mating. The next closest systems are the mating of half-sibs and that of double first cousins, followed by the mating of single first cousins. These systems are of theoretical interest and a full mathematical analysis of them and others is given by Fisher (1949). They are, however, elaborate systems and while of possible value to the plant or, more particularly, animal breeder they have never been found as a regular feature of wild species. Indeed the only system of regular inbreeding yet demonstrated in a wild species of animal is sib-mating: this occurs in the grass-mite *Pediculopsis* which is viviparous and in which the young mate before they are born (Cooper, 1937).

Assortative mating, which is the mating of phenotypically like individuals, is sometimes confused with inbreeding. Inbreeding, however, being the mating of relatives, always implies a particular relationship or set of relationships between mates in respect of all the genes that they carry and leads to homozygosis for all the genes in the genotype. Assortative mating does not do so. It will be effective only in relation to the genes affecting the character in respect of which the mating is assortative. Thus, for example, assortative mating in respect of leaf shape in plants might lead to homozygosis for genes affecting leaf-shape, but it would not do so for other genes affecting flower-colour or time of maturity or seed size, to mention but three other characters, except where the genes for these other characters were dragged along by linkage with the genes for the assortative character. The rise of homozygosis is also slowed down in assortative mating, even for the genes that it affects, if different genotypes share a common phenotype, which is never true of inbreeding. This may be illustrated by a simple example. In a population of plants varying in respect of a character governed by two alleles, full assortative mating would have exactly the same effect as selfing in producing homozygosis, provided the three genotypes GG, Gg and gg gave rise to three distinct phenotypes, for the mating of like with like in respect of phenotype then means that GG always mates with GG, Gg with Gg, and gg with gg. But suppose g is fully recessive to G. Then GG and Gg will share a common phenotype and while gg will always mate with gg, GG and Gg will mate indiscriminantly among themselves as a group. Starting with an F_2, the quarter of the population that is gg will give only gg offspring, but the remaining three-quarters will mate at random among themselves. This group consists of $\frac{1}{3}$ GG and $\frac{2}{3}$ Gg and hence has a gene frequency of $\frac{2}{3}$ G and $\frac{1}{3}$ g, with the result that its offspring will comprise $\frac{4}{9}$ GG, $\frac{4}{9}$ Gg and $\frac{1}{9}$ gg. These offspring will constitute $\frac{3}{4}$ of the next generation, the remaining being from gg parents and hence all gg. The next generation will thus consist of

$$\frac{4}{9} \times \frac{3}{4} = \frac{1}{3} \text{ GG;}$$

$$\frac{4}{9} \times \frac{3}{4} = \frac{1}{3} \text{ Gg;}$$

$$\frac{1}{4} + \left(\frac{1}{9} \times \frac{3}{4}\right) = \frac{1}{3} \text{ gg.}$$

In the following generation the population will comprise $\frac{3}{8}$ GG; $\frac{1}{4}$ Gg; $\frac{3}{8}$ gg, and two generations of this assortative mating will have raised homozygosis only to the level achieved by one generation of selfing or one generation of assortative mating where each genotype produces its own distinct phenotype. Where genes at two or more separate loci have similar effects, as in polygenic systems, assortative mating will be less effective still, for the same phenotype will be produced by different combinations of genes at the various loci. Furthermore, if non-heritable variation blurs the distinction between the phenotypes associated with the different genotypes and causes them to overlap, the efficacy of assortative mating will be reduced even further, whereas that of inbreeding proper would obviously be unaffected.

Assortative mating has, however, certain consequences which inbreeding does not produce. In particular, where two or more loci have similar effects on the character assortative mating tends to assemble alleles of like effect preferentially into the same breeding lines. Thus if G and H both tend to increase the expression of a character while their alleles g and h tend to reduce it. assortative mating tends to assemble G and H together into the same line of descent more often than would be expected by chance. The genes g and h would be similarly assembled preferentially, while the combinations G and h and g and H would appear less often than expected by chance. Inbreeding does not of itself have this effect.

So far in our discussion we have continued to assume that each individual, whether homozygous or heterozygous, makes an equal contribution to the next generation. It is a commonplace of genetics, however, that in species which do not normally inbreed, individuals produced by inbreeding are commonly less vigorous and less fertile than their non-inbred fellows. It is of interest, therefore, to examine the effects of such a reduced fitness in homozygotes on the progress of inbreeding. We will take for this purpose selfing as the simplest system of inbreeding and we will also make the simplifying assumption that the two classes of homozygote, GG and gg, are alike in their fitnesses. Let the population consist of a proportion a of homozygotes, and $b = 1 - a$ of heterozygotes and let the homozygotes produce $1 - s$ offspring for every 1 offspring of the heterozygotes. Under selfing, homozygotes breed true, but heterozygotes produce half homozygotes and half heterozygotes in their offspring. The homozygotes in the next generation will thus be $a(1-s) + \frac{1}{2}b$ and the heterozygotes will be $\frac{1}{2}b$. Since these add up to

$$a(1-s) + \tfrac{1}{2}b + \tfrac{1}{2}b = 1 - as$$

the proportion of homozygotes among the offspring will be

$$a' = \frac{a(1-s) + \tfrac{1}{2}b}{1 - as}$$

At equilibrium therefore $a' = \dfrac{a(1-s) + \tfrac{1}{2}b}{1 - as} = a$

giving $\qquad\qquad a - as + \tfrac{1}{2}b = a - a^2s$

and $\qquad\qquad\qquad as(1-a) = \tfrac{1}{2}b$

or $\qquad\qquad\qquad\quad abs = \tfrac{1}{2}b$

This relation is satisfied for any value of s if $b = 0$, i.e. when there are no heterozygotes in the population. But this is a trivial solution to the equation, as if no heterozygotes are present inbreeding has already run its course. It is thus of much more interest to examine the case when heterozygotes are still present, that is with $b \neq 0$. Then

$$as = \tfrac{1}{2} \quad \text{or} \quad a = \tfrac{1}{2s};$$

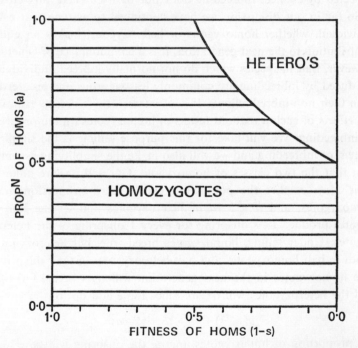

Fig. 2 The effects of inbreeding by selfing when homozygotes are at a disadvantage. When the fitness of homozygotes $(1-s)$ relative to that of heterozygotes (taken as 1) is greater than 0·5 complete homozygous is attained sooner or later. When $1-s$ is less than 0·5 complete homozygosis never results, the proportion of homozygotes obtained falling from 1 at $1-s-0·5$ to 0·5 when $1-s = 0$, i.e. when the homozygotes fail to make any contribution to the next generation.

By definition s can lie between 0, when the homozygotes are as fit as the heterozygotes, and 1, when the homozygotes leave no offspring. If $s<\frac{1}{2}$, a exceeds 1 and $b = 1-a$ must then be negative which is unreal. In other words, when $s<\frac{1}{2}$ there is no equilibrium except that achieved with $b = 0$. Given, therefore, $1-s>\frac{1}{2}$, i.e. that the homozygotes are at least half as fit as the heterozygotes, inbreeding will run its course until the population consists solely of homozygotes – though it should be noted that progress towards homozygosis will be slowed down when $s>0$.

If, however, s lies between $\frac{1}{2}$ and 1, a will lie correspondingly between 1 and $\frac{1}{2}$. To express this another way, if homozygotes are less than half as fit as heterozygotes, full homozygosis will never be attained no matter how long the inbreeding goes on. If the fitness of homozygotes, $1-s$, is only just below $\frac{1}{2}$, only a very small proportion of heterozygotes will persist; but if the fitness of homozygotes is 0, i.e. $s = 1$, half the population will continue generation after generation to consist of heterozygotes (see Fig. 2). Thus sufficiently heavy selection in favour of heterozygotes can prevent full homozygosis being attained and so vitiate the inbreeding.

Under selfing, a selective differential of at least 50% is necessary to prevent full homozygosis being attained. With less rigorous systems of inbreeding, smaller selective differentials will suffice. Sib-mating will fail to produce complete homozygosis if the fitness of the homozygotes falls only to 76% of that of the heterozygotes (i.e. $s = 0·24$) and with half-sib-mating a fall to only 81% (i.e. $s = 0·19$) will prevent it (Haymann and Mather, 1953).

Before leaving the mating system, a further point requires notice. If it is effective, inbreeding produces homozygosis and removes heterozygotes from the population. There is, however, no mating system which of itself will eliminate homozygotes and produce a population which consists generation after generation solely of heterozygotes. Let us make the extreme assumption that all homozygotes mate with other homozygotes of the opposite type, that is all their matings are GG × gg. Then a homozygote of the parent population will give nothing but heterozygotes in the next generation. But no matter what they mate with heterozygotes always give half homozygotes and half heterozygotes in their offspring. Thus the offspring generation will include $b' = a + \frac{1}{2}b$ heterozygotes and $a' = \frac{1}{2}b$ homozygotes. At equilibrium the proportion of heterozygotes must be constant and $b' = a + \frac{1}{2}b = b$. The maximum proportion of heterozygotes that can be maintained by adjustment of the mating system alone is thus $\frac{2}{3}$ and this only if each homozygote is restricted to mating with another of the opposite kind. We may note, too, that because GG must always mate with gg to give Gg offspring, the gene frequencies of G and g will be equal at equilibrium no matter how unequal they may be in the initial population. Where the frequencies of alleles G and g were equal, the population would consist of $\frac{1}{6}$ GG; $\frac{2}{3}$ Gg; $\frac{1}{6}$ gg and every homozygote would be restricted to $\frac{1}{6}$ of the population in obtaining a mate. Such a handicap, though severe, might not be regarded as crippling; but if we extend consideration to the case of genic variation at more than one locus, the same consideration must apply to homozygotes for the alleles at each of the loci and the restrictions would thus pile up until some of the genotypes would be virtually unable to find a mate. So not only is $\frac{2}{3}$ the maximum heterozygosis that can be obtained by adjustment of the mating system itself; even the attainment of this value for the many genes varying in a population would require an unrealistic restriction on mating. Random mating, with its maximum of a half heterozygosis, is therefore about as effective a system of outbreeding as can realistically be achieved. The importance of this point will appear when we come to consider the control of breeding systems in Chapter 6.

Levels of heterozygosity above $\frac{1}{2}$ can of course be achieved and maintained if there are more than two alleles at each locus; but this is as true with random mating as with any system of mating restricted so as to maximize heterozygosis. With three alleles all of equal frequency, $\frac{2}{3}$ of the population will be heterozygous under random mating; with four alleles the proportion will be $\frac{3}{4}$; and with n alleles it will be $n-1/n$.

Mutation

Adjustment of the mating system affects the distribution of the genes between homozygotes and heterozygotes, but save in special cases such as that discussed at the end of the last section, it does not of itself affect the frequencies with which the genes occur in the population. Mutation on the other hand does not directly affect the distribution between homozygotes and heterozygotes, but it does affect the gene frequencies. If G mutates to g at the rate μ_G, G will tend to be eliminated from the population unless there is some countervailing agent at work. Should g back-mutate to G at the rate of μ_g, equilibrium would be struck, other things being equal, when the increment lost by mutation from the frequency of G exactly balanced that added by back-mutation from g. When G occurs with frequency u, the increment lost by mutation will be $u\mu_G$, and since g occurs with frequency v, the increment gained by back-mutation will be $v\mu_g$. So equilibrium is achieved when

$$u\mu_G = v\mu_g$$

and the lower a gene's mutation rate by comparison with that of its allele the commoner it will be in the population at equilibrium.

Apart from the special class of so-called mutable genes which give rise to mosaicism in their carriers, mutation rates are low, being of the order of 10^{-5} or 10^{-6} or even lower. Achievement of equilibrium under mutation would thus be a slow process. Furthermore, these are the rates of mutation of wild-type or normal genes to their mutant alleles; back-mutation of mutant genes to their wild-type alleles would appear to be much rarer, as indeed is no more than to be expected if mutation is generally a greater or lesser breakdown of the normal structure of the gene on which its normal functioning depends. Given that back-mutation is rare, μ_G will be greater than μ_g and at equilibrium u will then be correspondingly lower than v. So if mutation was the only agency in operation, populations would come to be dominated by hypomorphic and amorphic mutant genes, with results so bizarre as to be scarcely imaginable. This is, however, manifestly not the case in nature: as we have already seen, though mutant genes, and hence mutant phenotypes, exist in populations each such gene is much rarer than its normal allele. Something must be opposing mutation and this is obviously selection, since the effects of the mutant genes are commonly deleterious to the individuals displaying them, the extreme case being where the gene is a lethal.

Where will equilibrium be struck between mutation and the selection which opposes it? Let us consider the simple case of a recessive mutant g, which reduces the fitness of its homozygous carriers, gg, to $1-s$, the fitness of GG and Gg being alike at 1. Let the frequencies of G and g be u and $v = 1-u$ respectively, and the mutation rate of G→ g be μ. Then u is reduced by $u\mu$ in each generation as a result of mutation. At the same time selection is tending to lower v. The frequencies of the three genotypes and their relative

Table 8   Effect of selection on gene frequency.

Genotype	GG	Gg	gg
Parental frequency	u^2	$2uv$	v^2
Fitness			
recessive gene	1	1	$1-s$
dominant gene	1	$1-s$	$1-t$
Offspring frequency			
recessive gene	u_r^2	$2u_r v_r$	v_r^2
dominant gene	u_d^2	$2u_d v_d$	v_d^2

where $u_r = u/(1-v^2 s)$, $\quad v_r = [v(1-vs)]/(1-v^2 s)$
$u_d = [u(1-vs)]/(1-2uvs-v^2 t)$, $\quad v_d = [v(1-us-vt)]/(1-2uvs-v^2 t)$

fitness are shown in Table 8, from which it will be seen that, leaving mutation aside for the moment, the gene frequencies among the offspring will be:

$$\text{for G, } u' = u/(1-v^2 s)$$
$$\text{for g, } v' = (v-v^2 s)/(1-v^2 s)$$

the divisor $1-v^2 s$ being introduced in order that $u'+v'$ shall equal 1.

Then v will diminish by a fraction $\dfrac{vs}{1-v^2 s}$ in every generation in the

absence of mutation. Bringing mutation also into account, however, gives
$$u' = (u - u\mu)/(1 - v^2 s)$$
and
$$v' = (v - v^2 s + u\mu)/(1 - v^2 s)$$

Then at equilibrium
$$v' = [v - v^2 s + (1-v)\mu]/(1 - v^2 s) = v$$
$$\text{giving } [v - v^2 s + (1-v)\mu - v + v^3 s]/(1 - v^2 s) = 0$$
$$\text{or } (1-v)\mu - v^2 s(1-v) = 0$$
$$\text{from which } v^2 = \frac{\mu}{s} \text{ and } v = \sqrt{\frac{\mu}{s}}$$

(except in the trivial case where $(1-v) = u = 0$ and the population is wholly gg). Since μ will generally be much smaller than s, it follows that v will be small and v^2, which is the frequency of the mutant phenotype, will be very small.

The case of a dominant mutant is a little more complicated since the heterozygote will then show the effects of the gene, with the result that its fitness will be reduced to $1-s$. The homozygous mutant can be expected to show the effects of the gene in an even more extreme form and correspondingly show a fitness further reduced to, say, $1-t$. These fitnesses are shown in the fifth line of Table 8. From them we find, when we also allow for mutation $G \to g$,
$$u' = [u^2 + uv(1-s) - u\mu]/[1 - 2uvs - v^2 t] = u$$
$$v' = [v^2(1-t) + uv(1-s) + u\mu]/[1 - 2uvs - v^2 t] = v$$
at equilibrium. Then from $v' = v$
$$v - uvs - v^2 t + u\mu = v - 2uv^2 s - v^3 t$$
which reduces to
$$vs - 2v^2 s + v^2 t - \mu = 0$$
and
$$v = \frac{\mu}{s - v(2s - vt)}$$

Since μ will be generally several orders of magnitude less than s, v will be small. We can, therefore, conveniently neglect the terms in v in the denominator of the fraction and write $v = \frac{\mu}{s}$

Since v is small, v^2 will be very small and the gg phenotype will be virtually non-existent in the population. The mutant phenotype is, however, shown by Gg which appears with a frequency $2uv$. Since v is small, u approximates to 1 and we write the expected frequency of mutants in the population as

$$2v = \frac{2\mu}{s}$$

The best documented case of balance between mutation and selection is that of the dominant gene which determines achondroplasia, a form of dwarfing, in man. Mørch (1941) records that out of 94 075 children born in certain Copenhagen maternity hospitals 10 were achondroplasiacs, but of these 10 only two had an achondroplasiac parent as would be expected with a dominant gene. The other eight children must therefore be regarded as carrying newly mutated achondroplasiac genes. Since each child carries two genes and only one would be newly mutated, we find the mutation rate as

$$\mu = 8/(2 \times 94\,075) = 43 \times 10^6$$

Now Mørch traced 108 achondroplasiacs in the Danish population and these had only 27 children between them, whereas 457 normal sibs of these achondroplasiac dwarves had 582 children. The relative fitness of the dwarves, taking that of their normal sibs as unity, is thus

$$1 - s = \frac{27}{108} \div \frac{528}{457} = 0\cdot 2$$

giving $\qquad\qquad s = 1 - 0\cdot 2 = 0\cdot 8$

We note immediately that the loss of fitness due to this gene is $0\cdot 8$ which equals the proportion of achondroplasiac births that must be ascribed to mutation, so suggesting that selection and mutation are in balance. This we can test further because at equilibrium the frequency of affected individuals for the population should be

$$2v = \frac{2\mu}{s} = \frac{2 \times 43}{10^6 \times 0\cdot 8} = 108 \times 10^{-6}$$

Among the 94 075 children Mørch observed, 10 were in fact affected, so giving as the frequency in the population

$$\frac{10}{94\ 075} = 106 \times 10^{-6}$$

which agrees closely with expectation. Since an achondroplasiac carries one mutant gene and one normal, the frequency of the gene in the population is

$$v = \frac{\mu}{s} = 53 \times 10^{-6}$$

To recapitulate, a gene is at equilibrium in the population when $v = \sqrt{\mu/s}$ if it is a recessive mutant and where $v = \mu/s$ if it is a dominant; and affected individuals are present in the population with frequency μ/s for recessives and $2\mu/s$ for dominants. The incidence of affected individuals can thus be raised in either of two ways, by increasing μ or by decreasing s. Exposure to ionizing radiations increases μ and the growing use of such radiations in medicine, industry and agriculture, with consequent growing exposure of people to them, has raised the question of the extent to which we are thereby raising the prospective incidence of genetic abnormalities in our populations. This has been discussed in a number of official publications, in Great Britain in a Medical Research Council Report (1956) which was followed by legislation designed to limit exposure to ionizing radiations to a level acceptable in that it neither deprives us of the valuable uses of these radiations, nor opens up a risk of serious harm either to those exposed to the radiations or to their descendants. Chemical substances that have a mutagenic effect are now coming under similar scrutiny. It should be noted that the effects of an increased mutation rate in raising the incidence of disabilities in the population would be felt relatively quickly in the case of dominant mutation, but only very slowly with recessives, where indeed the incidence of disability would have moved only about half way towards its new equilibrium in fifty generations.

The incidence of genetic disability will increase also as s is decreased, and this will come about in so far as medical science is successful in treating disabilities springing from the action of

D

mutant genes. Thus, for example, the development of insulin treatment for diabetes melitus must be raising the frequency of genes which bring this disease about, for it has clearly raised the prospect of diabetics reproducing. Pyloric stenosis, whose genetic causation is hardly to be questioned even though the detailed pattern has yet to be finally established, was fatal to babies born with it until some forty years ago, an operative treatment was introduced which allows them to survive and grow up normally. Children successfully treated in the earlier days of the operation are now themselves parents and it is clear that the incidence of the disability is markedly higher among their children than it is in the general population (Carter 1961). Again there can be no doubt that genes responsible for pyloric stenosis must be increasing in frequency. As with increase in μ, decrease in s will produce a more rapid increase in disabilities due to dominant genes than in disabilities due to recessives. In general the increase will be somewhat slower when it depends on reduction in s than when it depends on increase in μ.

Immigration, or the accruel to the population of individuals from an external source, has the same effect as mutation in changing the gene frequency when that of the immigrants differs from that of the population. It can lead to an equilibrium with selection, and can be dealt with algebraically in much the same way as mutation. It differs from mutation, however, in two important respects. First, it is not restricted to the same low rate as mutation and so can maintain in the population higher frequencies of the immigrant allele and hence of aberrant individuals even where a balance is struck with opposing forces of selection. Second, whereas mutation is normally a change in an individual gene, genes at other loci being unaffected by it, immigration can be altering the gene frequencies of many loci simultaneously since the immigrants may differ from the recipient population in gene frequency at many loci.

In our discussion of the action of selection in opposing the effects of inbreeding and mutation we have so far assumed differences in fitness to be unconditional in the sense that the fitness of an individual was independent of the other individuals present in the population and, therefore, of the genetical structure of the population. Much

selection must, however, come about as a result of competition among the individuals of the population and its impact will, therefore, vary with the composition of the population (Mather 1969). Suppose we have two types, A and B, present in the population with frequencies x and $y = 1 - x$, A having an advantage over B when the two come into competition. An individual A will meet competition from another A in a proportion x of cases and its fitness will then be 1; but in y of cases it will meet B in competition and, having the advantage, its fitness will be increased to $1 + k$. Similarly a B individual will meet competition from its own kind and hence have unit fitness in y of cases, but meet competition from A and have its fitness reduced to $1 - k$ in x of cases. Then the average fitness of type A will be
$$x + y(1 + k) = 1 + yk$$
and that of type B will be $y + x(1 - k) = 1 - xk$

The total contribution of A individuals to the next generation will thus be $x(1 + yk)$ and of B's will be $y(1 - xk)$, the joint contribution thus being 1. The value of k will obviously depend on the intensity of the competition and therefore in general on the relative density of the population. The selection is thus density-dependent.

Let us investigate the equilibrium such competitive selection will give with mutation in the simple case of a recessive mutant gene. Since GG and Gg have the same phenotype in such a case they jointly compose type A with a frequency of $u^2 + 2uv = 1 - v^2$ and an average fitness of $1 + kv^2$. Type B will consist of the gg individuals with a frequency of v^2 and an average fitness of $1 - k(1 - v^2)$. The mutational increment to v will be $u\mu$ as before. Then at equilibrium
$$u' = (u^2 + uv)(1 + kv^2) - u\mu = u$$
and
$$v' = uv(1 + kv^2) + v^2[1 - k(1 - v^2)] + u\mu = v$$
there being no divisor necessary to make $u' + v' = 1$.

Then from $u' = u$ we find $u + uv^2k - u\mu = u$

giving
$$v^2 = \frac{\mu}{k} \text{ and } v = \sqrt{\frac{\mu}{k}}$$

just as in the earlier calculation. Thus the equilibrium is independent of whether the differences in fitness are unconditional or are dependent on competition. One point may, however, be noted. Since the value of k, measuring the intensity of selection, will vary with for example the density of the population, the point of equilibrium will change as the circumstances affecting the population and hence the intensity of competition change.

Selection

We have seen that both the rise of homozygosis under inbreeding and change in gene frequencies by mutation can be opposed by an appropriate action of selection and an equilibrium so attained. Selection can clearly have a variety of effects and we must now turn to take a more general look at them.

Let us take the fitness of GG as our point of reference and designate it as unity. Then, following the notation used for the case of the dominant gene, the fitness of Gg is $1-s$ and that of gg is $1-t$. If s and t are positive ($+ve$), Gg and gg have lower fitnesses than GG; but if s is negative ($-ve$) Gg is of higher fitness than GG, and if t is $-ve$ the same is true of gg. Clearly neither s nor t can be greater than 1 for this would imply $-ve$ fitness, though in principle there is no limit to the size of the $-ve$ value they may take when the fitness becomes greater than unity. It should be noted that taking the fitness of GG as 1 is no more than using it as a convenient point of reference: it is a relative value only and does not imply that each GG individual actually contributes one individual offspring (or a half share of two individuals) to the next generation. The contribution of any parent individual to the next generation of the population will obviously vary according to whether the population is increasing, decreasing, or stable in size and according to the contribution made by individuals of other genotypes.

Now, with the fitnesses as set out in the bottom line of Table 8, the proportion of G genes in the next generation will be

$$u' = [u^2 + uv(1-s)]/[1 - 2uvs - v^2t] = [u - uvs]/[1 - 2uvs - v^2t]$$

and $v' = [v^2(1-t) + uv(1-s)]/[1 - 2uvs - v^2t]$

$$= [v - v(vt + us)]/[1 - 2uv - v^2t]$$

the denominator being included to make $u' + v' = u + v = 1$.

Then, consequent on the action of selection, the frequency of G changes by

$$\Delta u = u' - u = [(u - uvs)/(1 - 2uvs - v^2t)] - u$$
$$= [(u - uvs) - (u - 2u^2vs - uv^2t)]/[1 - 2uv - v^2t]$$
$$= uv[us - v(s - t)]/[1 - 2uvs - v^2t]$$

Now the denominator must be $+ve$ except when no progeny at all are left in the next generation. The general properties of Δu will thus depend on the numerator.

When $uv[us - v(s - t)]$ is $+ve$, u will be increasing;

When $uv[us - v(s - t)]$ is $-ve$, u will be decreasing

and when $uv[us - v(s - t)] = 0$, the population will be at equilibrium. We may note in passing that equilibrium is achieved when $u = 0$ (i.e. the genes G is not present in it) or when $v = 0$ (i.e. the gene g is not present), but these are trivial solutions to the equation and the interesting equilibria will be defined by $us - v(s - t) = 0$. Also since uv must be $+ve$, the sign of Δu in the absence of equilibrium will depend on the sign of $us - v(s - t)$.

When s and t are both $+ve$, and $s < t$, $us - v(s - t)$ must be $+ve$ with u therefore increasing until $v = 0$. When s and t are both $-ve$, and $-s < -t$, $uv - v(s - t)$ must be $-ve$ with u therefore decreasing until $u = 0$. In these cases therefore one allele will increase in frequency until it has entirely replaced the other (in the absence of mutation or immigration) G being the successful allele when $us > -v(s - t)$, and g when $us < -v(s - t)$.

In other cases the sign of $us - v(s - t)$ will depend also on the values of u and v and equilibrium is possible when

$$us - v(s - t) = 0 \text{ or } us = v(s - t).$$

This condition can be satisfied at appropriate values of u and v when s is $+ve$ and t is $-ve$, or when s is $-ve$ and t is $+ve$, or if both are of the same sign when $-s < -t$. A little reflection will show that this is equivalent to saying that equilibrium is possible when the fitness of the heterozygotes, Gg, is either less than those of both GG and gg, or greater than those of both GG and gg. In all other cases one of the alleles will oust the other completely from the population.

Is there any general difference between the two equilibria, achieved when the heterozygote is more fit than both homozygotes, or less fit than both? Let $\hat{u}$ and $\hat{v}$ be the equilibrium values of u and v, so that $\hat{u}s = \hat{v}(s-t)$ and let any $u = \hat{u}+\delta u$ and $v = \hat{v}-\delta u$.

Then for any values of u and v, the change in u is

$$us - v(s-t) = \hat{u}s - \hat{v}(s-t) + \delta us - (-\delta u)(s-t)$$
$$= \delta u(2s-t)$$

As we have seen above, for equilibrium with Gg less fit than both homozygotes s is $+ve$ and $t<s$. Then $2s-t$ must be $+ve$. So if u is less than $\hat{u}$, δu is $-ve$. So Δu is $-ve$ and u will get smaller. Similarly if $u>\hat{u}$ and δu is $+ve$, Δu will be $+ve$ and u will get larger. In other words when not already at their equilibrium values the gene frequencies will move away from the equilibrium until either G or g, whichever has the smaller frequency is eliminated. It must thus be an unstable transient equilibrium and as such trivial for understanding population structure.

But, for equilibrium with Gg at an advantage over both homozygotes s is $-ve$ and $-t<-s$. Then $2s-t$ must be $-ve$ and if u is less than $\hat{u}$, δu is also $-ve$ so that $u = \delta u(2s-t)$ is $+ve$. Thus u will increase until it attains $\hat{u}$. Similarly with $u>\hat{u}$, δu is $+ve$ and Δu will be $-ve$ with the result that again u decreases until it attains $\hat{u}$. Departure of the gene frequencies from the equilibrium is thus self-correcting, the equilibrium tends to be restored after any departure from it, and it is consequently said to be a stable equilibrium. This is therefore an important type of equilibrium.

If we are considering only these equilibria, and not the general action of selection, it is easier to set the fitness of Gg at unity, calling the fitness of GG $1-s_G$ and that of gg $1-s_g$. Then it is easy to see that

$$\frac{u'}{v'} = \frac{u^2(1-s_G)+uv}{v^2(1-s_g)+uv} = \frac{u(1-us_G)}{v(1-us_g)} = \frac{u}{v} \text{ at equilibrium.}$$

The condition for equilibrium is thus $us_G = vs_g$ or $\dfrac{u}{v} = \dfrac{s_g}{s_G}$, and if both s_G and s_g are $+ve$ (i.e. GG and gg are of less fitness than Gg) the equilibrium is stable, whereas if s_G and s_g are both $-ve$ (i.e. GG and gg are of greater fitness than Gg) the equilibrium is unstable.

Thus stable equilibrium depends on heterozygous advantage, as we saw from our earlier and more general consideration. We may note that us_G is equivalent to our earlier us, and us_g to $v(s-t)$ though s_G is of opposite sign to s.

An excellent example of such polymorphism is afforded by the gene for sickle-cell, or S-type haemoglobin in man (Allison 1955). The normal gene, which we may denote as H, leads to the production of normal haemoglobin, while the mutant, h, leads to a haemoglobin differing from the normal by the substitution of valine for glutamic acid at a particular site in it. This S haemoglobin is less efficient than the normal in oxygen transport, but on the other hand it confers resistance to malignant tertian malaria. In consequence hh homozygotes, which produce only S haemoglobin, are resistant to this malaria but suffer from sickle-cell anaemia which causes distorted development, commonly leads to early death and reduces fitness to an estimated 25% of normal. HH homozygotes have only normal haemoglobin and so do not suffer from the anaemia. They are, however, susceptible to the malaria. Heterozygotes, Hh, enjoy the best of both worlds since they carry both types of haemoglobin in nearly equal amounts, the normal haemoglobin being sufficient to avoid the anaemia and the S-type to protect against the malaria. In malarial country, therefore, HH individuals are at a disadvantage by comparison with the heterozygotes, the degree of disadvantage depending on the morbidity of the normals from malaria. The sickle-cell gene, which we have called h, is found in African negroes and tribes inhabiting malarial environments are polymorphic for it. In extreme cases, as many as 4% or 5% of children may be born with the anaemia, implying a gene frequency of 0·8H and 0·2h. Taking the fitness of anaemics as $1 - s_h = 0·25$, the equilibrium relation in either its simpler final form $us_H = vs_h$ or the more general $us = v(s-t)$ would then suggest that the morbidity rate of normals from malaria is 0·18. Lower frequencies of h imply lesser risks from malaria, and the gene is indeed rare in tribes inhabiting non-malarial environments. As would be expected, the polymorphism is observed where both homozygotes are less fit than the heterozygote and it is absent where only one homozygote is less fit, because in the absence of malaria normals suffer no morbidity from it. We might observe in passing that the way to remove h, and with it sickle-cell anaemia, from the population is thus curiously enough to eliminate the risk of malaria. Search has been made for heterozygous advantages in other polymorphisms in man. In some cases, including the ABO blood groups and taste-blindness, relations have been established of the phenotypic classes to the incidence of certain diseases (see Mather 1964) but no hetero-zygous advantage has yet been demonstrated.

Man does, however, also appear to afford an example of the heterozygote being at a disadvantage by comparison with both homozygotes. The Rhesus blood group gene is complex, but for the purpose of considering haemolytic disease of the newborn it may be regarded as effectively comprising two alleles, Rh+ and Rh−, the latter being recessive. If an Rh− mother carries an Rh+ foetus she is liable to make antibodies against the Rh+ antigen. These anti-bodies can enter the circulation of later foetuses and if any of these is again Rh+ the reaction may bring about haemolytic disease. Now an Rh+ foetus in an Rh− mother must be heterozygous. Homozygous Rh+ foetuses can be borne only by Rh+ mothers, and the Rh− foetuses must come from Rh− mothers. So only the heterozygotes are subject to haemolytic disease. It is still something of a problem, therefore, why human populations throughout the world are still polymorphic for Rh+ and Rh−. Either they are not in equilibrium and are moving, as yet undetectedly, towards the elimination of the rarer allele, Rh−, or there are other forces of selection at work, about which we may speculate but have yet to demonstrate, which alter the picture as it appears to us at present.

Competitive and frequency-dependent selection

The equilibria which can be produced in a population by uncondi-tional selection can also arise from competitive selection. Let the relations of the three genotypes be such that GG in competition with Gg has a fitness of $1+k_a$, and in competition with gg a fitness of $1+k_c$; Gg in competition with GG, a fitness of $1-k_a$ and in com-petition with gg a fitness of $1+k_b$, while gg in competition with GG has a fitness of $1-k_c$ and with Gg one of $1-k_b$. These relations are shown diagrammatically in Fig. 3. When k_a, k_b and k_c are $+ve$, GG is the fittest genotype and gg the least fit, whereas if they are all $-ve$ the order of fitness is reversed.

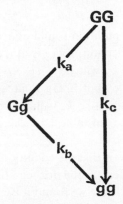

Fig. 3 The competitive differentials among the three genotypes possible with two alleles, G and g. In each case $1+k$ is the fitness of the genotype from which the arrow originates and $1-k$ that of the genotype to which the arrow points. Thus when GG and Gg are in competition GG has a fitness of $1+k_a$ and Gg a fitness of $1-k_a$, etc.

GG will compete with itself in u^2 of cases, with Gg in $2uv$ of cases and with gg in v^2 of cases. Its average fitness is therefore

$$u^2 + 2uv(1+k_a) + v^2(1+k_c) = 1 + 2uvk_a + v^2k_c$$

The fitnesses of Gg and gg will similarly be $1 - u^2k_a + v^2k_b$ and $1 - 2uvk_b - v^2k_a$ respectively.

Then the output of gametes will be

G: $\quad u' = u^2(1 + 2uvk_a + v^2k_c) + uv(1 - u^2k_a + v^2k_b)$
$\qquad\quad = u(1 + u^2vk_a + v^3k_b + uv^2k_c)$

g: $\quad v' = u^2(1 - 2uvk_b - u^2k_c) + uv(1 - u^2k_u + v^2k_b)$
$\qquad\quad = v(1 - u^3k_a - uv^2k_b - u^2vk_c)$

At equilibrium $\quad u' = u(1 + u^2vk_a + v^3k_b + uv^2k_c) = u$

from which $u^2k_a + v^2k_b + uvk_c = 0$ where neither u nor v equals 0.

Any population which satisfies this condition will be in equilibrium under selection alone.

If the effects of competition on fitness are simply additive in the sense that $k_a + k_b = k_c$ the condition for equilibrium becomes

$$(u^2 + uv)k_a + (v^2 + uv)k_b = uk_a + vk_b = 0$$
$$\text{or } uk_a = -vk_b$$

This is similar to the condition found above in our general treatment of unconditional selection: if Gg is at a competitive advantage over both homozygotes k_a will be $-ve$, k_b will be $+ve$, and the equilibrium will be stable; but if Gg is at a disadvantage against both homozgotes k_a will be $+ve$, k_b will be $-ve$ and the equilibrium though attainable will be unstable and therefore even if attained will be transient.

We should note, however, that the condition for equilibrium under competitive selection offers scope for equilibrium to be reached under conditions beyond those we found for unconditional selection since k_c need not equal $k_a + k_b$. Wherever

$$k_c = -\left(\frac{u}{v}k_a + \frac{v}{u}k_b\right)$$

equilibrium can be attained and in particular even if k_a and k_b are of the same sign equilibrium is attainable given that k_c is sufficiently large and of the opposite sign. Thus even where GG is superior in competition with Gg and Gg with gg, equilibrium is still attainable provided that in its turn gg has a sufficiently large advantage in direct competition with GG. This is perhaps on the face of it an unlikely situation, but one that nevertheless cannot be ruled out as a possibility in nature.

The values of k_a etc. will vary with the frequencies in which competitive situations occur between individuals and their intensity when they do arise. These in their turn must depend on the density of individuals on the ground covered by the population. Competitive selection will therefore be conditional and density-dependent (Mather 1969). In a sense, too, it will depend on the values of u and v, the gene frequencies, since the outcome of competition depends on the genotypes of the competing individuals whose frequencies, and with them the frequencies of the different competitive situations, depend on u and v. Competitive selection is therefore frequency-dependent as well as density-dependent. Frequency-dependent selection can, however, arise and lead to equilibrium in other ways. Consider the situation in a plant with an incompatibility mechanism of the type well known in for example *Oenothera organensis* (see Lewis 1954). Here the incompatibility reaction is controlled by a long series of alleles at the S locus, the genes acting in such a way that a pollen tube carrying any given allele cannot successfully grow down a style which carries that same allele, and thereby achieve fertilization (Fig. 14 on p.112). Individuals, therefore, can never be homozygous for an S allele, and the success of a pollen grain in finding a style on which it can be effective will be the greater the rarer the allele which it itself carries. We shall have occasion to refer to such incompatibility again later (Chapter 6), but it is sufficient for our present purpose to note that the simplest possible system of this type involves three alleles S_1, S_2 and S_3. Individuals will therefore be of three types S_1S_2, S_1S_3, and S_2S_3. Let the frequencies of these genotypes be p, q and r respectively, where $p+q+r = 1$. Then the frequency of pollen carrying S_1 will be $\frac{1}{2}(p+q)$, of pollen carrying S_2 will be $\frac{1}{2}(p+r)$ and of pollen carrying S_3 will be $\frac{1}{2}(q+r)$. But S_1 pollen cannot grow down the styles of S_1S_2 and S_1S_3 plants. Assuming that S_1 pollen is distributed at random to the styles of the three types of plants it will be successful in only $r = (1-p-q)$ of them. S_1 pollen will therefore be successful in $\frac{1}{2}r(p+q)$ of fertilizations. Similarly S_2 and S_3 pollen will be successful in $\frac{1}{2}q(p+r)$ and $\frac{1}{2}p(q+r)$ of fertilizations. S_1 pollen in an S_2S_3 individual will give S_1S_2 and S_1S_3 individuals in equal numbers (Table 9) and we find therefore that the composition of the next generation will be

$$S_1S_2: p' = pq+pr+2qr/4k$$
$$S_1S_3: q' = pq+2pr+qr/4k$$
$$S_2S_3: r' = 2pq+pr+qr/4k$$

where $k = pq+pr+qr$ is included in the denominator to make $p'+q'+r' = 1$.

Then $p'-q' = \dfrac{1}{4k}(qr-pr) = \dfrac{r}{4k}(q-p) = p-q$ at equilibrium.

Since r and k must both be $+ve$, this relation is satisfied only when

Table 9   Frequency dependent selection for incompatibility genes

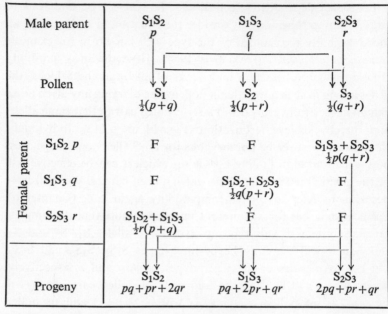

Male parent	S_1S_2 p	S_1S_3 q	S_2S_3 r
Pollen	S_1 $\frac{1}{2}(p+q)$	S_2 $\frac{1}{2}(p+r)$	S_3 $\frac{1}{2}(q+r)$
$S_1S_2\ p$	F	F	$S_1S_3+S_2S_3$ $\frac{1}{2}p(q+r)$
$S_1S_3\ q$	F	$S_1S_2+S_2S_3$ $\frac{1}{2}q(p+r)$	F
$S_2S_3\ r$	$S_1S_2+S_1S_3$ $\frac{1}{2}r(p+q)$	F	F
Progeny	S_1S_2 $pq+pr+2qr$	S_1S_3 $pq+2pr+qr$	S_2S_3 $2pq+pr+qr$

$p = q$. By extension it can be shown that $p = q = r$ at equilibrium, which is thus achieved when all these genotypes, S_1S_2, S_1S_3, and S_2S_3, are equally common, each forming a third of the population. By derivation the frequencies of the three alleles, S_1, S_2 and S_3 are also all equal at $\frac{1}{3}$.

Since when $q>p$, $p'>q'$ and *vice versa*, an excess of say S_1 in one generation will lead to a shortage, albeit a smaller shortage, of this allele in the next. The equilibrium is thus stable and will be approached by a series of diminishing oscillations round the equilibrium values.

Thus in the control of incompatibility an allele, because of its action, is at a selective advantage when its frequency is low, and at a disadvantage when its frequency is high. The selection is frequency-dependent and equilibrium is achieved because it is so. We should note, however, that not all frequency-dependent selection will lead to an equilibrium of this kind, for if the selective advantage of an allele becomes greater as that allele becomes commoner it will inevitably lead to extinction of other alleles, barring mutation and immigration.

The operation of frequency-dependent selection is easy to recognize in the genetical control of breeding system, but it is also known to occur in other cases. Huang, Singh and Kojima (1971) have demonstrated that it occurs in relation to the alleles at the esterase-6 locus in *Drosophila melanogaster*. These flies appear to secrete or excrete into their substrate materials which have a deleterious effect in competing individuals, the deleteriousness being greater for individuals of the same genotype than others. Thus the commonest genotype is at the greatest disadvantage, selection is frequency-dependent and a polymorphism ensues for the alleles at this locus. It has been observed, too (Ehrman and Petit, 1968) that individuals of *Drosophila willistoni* take part in mating disproportionately often when they are of a relatively rare type, again producing frequency-dependent selection. Finally Cook (1971) records observations on the predatory behaviour of birds which indicate that when the prey is polymorphic, rarer morphs are taken less frequently than even their rarity would suggest, and commoner morphs correspondingly more often than their frequency would suggest. Thus a rarer gene producing a rare morph would, because of its rarity, have an advantage over a commoner gene producing a commoner morph. Selection is again frequency-dependent of a kind that would lead to a polymorphic equilibrium, though there is some evidence (Allen, 1972) to suggest that in a very dense population this advantage is lost and even reversed. Thus frequency-dependent selection may itself be density-dependent, so illustrating yet again the complexity and intricacy of the action of selective forces.

We have seen that selection can operate in a wide variety of ways. It can lead to the fixation of one allele and the extinction of another. It can affect and even vitiate the action of inbreeding, and it can strike a balance with the action of mutation and immigration. It can also produce polymorphic equilibria by its own action where this favours alleles differentially in special combinations, special frequencies or special circumstances. We have seen how selection operates when it is unconditional, conditional and density-dependent, and frequency-dependent. The treatment has aimed at no more than illustrating the versatility of selection in its operation and the methods of assessing its consequences. Fuller discussions of various aspects of the action of selection and the mathematical analysis of its consequences will be found in, for example, Li (1967), Crow and Kimura (1970) and Cook (1971).

Genetic load

The occurrence of these various types of gene variation, whether rare deleterious mutants or polymorphisms, implies that there exists in the population individuals whose fitness is reduced to various degrees. The average fitness of the individuals who together compose the population is thus reduced, and this reduction is known as the genetic load of the population. Load is defined as the proportionate decrease in average fitness relative to that of the fittest or optimal genotype thus

$$L = (W_m - \overline{W})/W_m$$

where L is the genetic load, W_m the fitness of the fittest genotype and $\overline{W}$ the average fitness of all individuals.

Load may arise for a variety of reasons, mutation, migration, incompatibility of mother and foetus as encountered with the Rh gene, non-disjunction of chromosomes, hybridization of otherwise discrete populations, polymorphism and so on; but only two of the types of load need detain us, that due to deleterious mutations (the mutational load) and that due to segregation in polymorphic populations, where these are maintained by heterozygous advantage (the segregational load). Taking mutational load first, we have seen that the incidence of affected individuals in a population where the mutant is at equilibrium under the opposing forces of mutation and selection is μ/s for recessive mutant genes and $2\mu/s$ for dominant. Each affected individual represents a loss of fitness of s. So multiplying the frequency of affected individuals by their loss of fitness we find that the load in the population is $(\mu/s) \times s = \mu$ for recessive genes and $(2\mu/s) \times s = 2\mu$ for dominants. The load is thus directly

proportional to the mutation rate of the gene and independent of
s – at first glance a surprising result, but one which becomes less
so when we remember that the more unfitting the gene the lower
the frequency of affected individuals in the population. Taking any
single locus, the load its mutation produces is thus very small, of
the order of 10^{-5} or 10^{-6}. But bringing into account all the many
loci which can mutate to produce alleles of deleterious effect, the
load will be much higher because it will be approximately the sum
of all the mutation rates. Thus with 1000 such loci the load might
be of the order of 10^{-3} to 10^{-2} or $0 \cdot 1 \% - 1 \%$. With 10 000 such loci
it would be some 10 times as high again.

In turning next to the segregational load, we will use the notation
of page 46, taking the fitness of the heterozygote Gg as unity and
that of GG and gg as $1 - s_G$ and $1 - s_g$ respectively. Then with
heterozygous advantages, that is both s_G and s_g +ve, and the
frequencies $u^2 : 2uv : v^2$ for the three genotypes, at equilibrium the
population will lose fitness to the extent of $u^2 s_G + v^2 s_g$ because of
the polymorphism. Now we saw earlier that, at equilibrium, $u s_G = v s_g$
which since $v = 1 - u$, reduces to

$$u = \frac{g_s}{s_G + s_g}$$

and similarly

$$v = \frac{s_G}{s_G + s_g}$$

Thus the loss of fitness is

$$\frac{s_g^2 s_G}{(s_G + s_g)^2} + \frac{s_g s_G^2}{(s_G + s_g)^2} = \frac{s_G s_g}{s_G + s_g}$$

We saw (p. 47) that with the sickle-cell polymorphism s_g was of the
order of $0 \cdot 75$ and in malarial regions s_G could be nearly as high as $0 \cdot 2$.
This would lead to a load of $\dfrac{0 \cdot 2 \times 0 \cdot 75}{0 \cdot 2 + 0 \cdot 75} = 0 \cdot 16$. Thus this single
polymorphism could give rise to a load as great as that from a very
large number of loci mutating to produce deleterious genes. This
particular polymorphism involves what are no doubt unusually high
selective disadvantages, but even so it is clear that unless poly-

morphism is rare the genetic load, as it is defined and measured, will spring much more from polymorphic segregation than from mutation, and furthermore it could be high. The finding that in mammals and arthropods (see p. 20) polymorphism for biochemical traits can be common, and indeed may well be displayed by one third or more of the relevant loci, has therefore raised the question of whether a population could carry the load of unfitness that so many polymorphisms maintained by such selection must impose and still reproduce sufficiently well to survive. The alternative view is that the polymorphisms, or at any rate many of them, spring from selectively neutral allelic differences and exist and change in the population by random drift of the gene frequencies (see p. 82). Such a view must be in the nature of a last resort, to be adopted only if an acceptable interpretation cannot be found in terms of selection, for it is clear that some polymorphisms like sickle-cell in man and industrial melanism in insects (see Cook, 1971) involve selective differences, and there is also reason to think that the biochemical polymorphisms in man do so, albeit no doubt of lesser intensity (Harris, 1971), as do those in *Drosophila equinonalis* (Ayala *et al*, 1972) and in the plant *Avena barbata* (Clegg and Allard, 1972; Allard *et al.*, 1972). Furthermore, the very idea of genetic load as applied in this way is open to a number of serious objections.

As we have seen, load is defined by reference to the most fit genotype. With n polymorphisms, each maintained by heterozygous advantage, this will be the multiple heterozygote. But if in respect of any locus half the individuals of the populations are heterozygous, which is the maximum for a pair of alleles, a proportion of $(\frac{1}{2})^n$ of individuals will be simultaneously heterozygous for all n loci. With $n = 10$, this is about one in a thousand, with $n = 20$ it is about one in a million and with $n = 40$ one in a million million. Evidently a perfectly adequate level of fitness can be achieved by individuals not at the theoretical maximum and since these will constitute the bulk of the population they set the standard of comparison. So the load as formally defined must be misleadingly high (see Wallace, 1970).

Furthermore it is assumed in load calculations that selective disadvantages are unconditional ('hard selection' in Wallace's terminology) and this is by no means always, or even generally, the case. With conditional selection (Wallace's 'soft selection'), the idea of load takes on a different appearance. Where selection is frequency-dependent, all genotypes are equally fit when they are present in equilibrium proportions, differing in fitness only as they depart from the equilibrium. So such selection produces no load in the formal sense except when departing from equilibrium, and then it is self-correcting. When we turn to competitive selection, with its density-dependent relation, we bring in still further considerations (Mather, 1969).

No population exists in a biological vacuum. True, its individuals must be capable of surviving and leaving on average at least one offspring each under the prevailing physical conditions of temperature water supply, etc. (or of course two offspring if each of these must have two parents). But in addition and more immediately important save under exceptional conditions, each individual must be capable of surviving and leaving offspring while under intense competition from its fellow individuals as well as those of other species occupying the same habitat and, to judge by the number of seed produced by most plants or fertilized eggs produced by most animals, all able in the abstract of leaving considerably more than one or two offspring. A given habitat will, however, maintain only a certain totality of individuals and competition for the limiting resources, such as light or a nutrient, will play a major part in determining which individuals leave offspring and in what numbers.

Competition capable of resulting in a genetical adjustment or readjustment of a population must be directly or indirectly between the member individuals of that population. Thus fitness is essentially a relative and conditional thing: the success or otherwise of an individual in competition does not depend just on itself and its own genotype or on its capabilities as judged by comparison with some other individual which may indeed be rare or even absent from the population but whose genotype theoretically endows it with a maximal fitness. On the contrary, success or lack of it in competition is determined by the properties and genotypes of the individuals

E

actually competing. It is thus by comparison solely with the rest of the actual population that we must judge an individual. Individuals may survive and reproduce perfectly successfully because there is nothing to oust them in competition, though the moment that a superior genotype arose or entered the population they would become at a disadvantage. Load as formally defined and measured cannot give a true picture of a population and its prospects. Furthermore competitive selection is relative in yet another way. It will commonly be density-dependent and if for any reason a population should decline in numbers we should expect the pressure of competitive selection (as represented by the value of k in our analysis) to decline, with the result that disadvantages would be lessened and the decline in numbers tend to self-regulation. In the same way an increase in numbers would tend to strengthen selective pressures and again be self-regulating.

A population will tend to cash in on an increase in fitness (whether this be by some amelioration of the environment or by the introduction of fitter genotypes) by an increase in its numbers through a rise above the mere replacement rate of the average number of offspring successfully left in the next generation. Similarly a worsening of the environment will tend to lead to a decrease in numbers with a restoration of the erstwhile level if and when new and fitter genotypes arise. Thus a reduction in the strength of competition may be marked by a rise in numbers accompanied by an increase in the variation detectable in the population as was indeed observed by Ford and Ford (1930) in the butterfly *Melitaea aurina*. Similarly a decline in numbers, marking more intense competition, would be accompanied by a reduction in detectable variability. This relation of selective pressure and population size is one that is commonly overlooked in genetical discussions of selective action where population size is tacitly assumed to be constant and selection generally unconditional. The relation is complex and has yet to be explored adequately, though a valuable start has been made on its consequences for the concept of genetic load by Wallace (1970).

A special aspect of the problem of genetic load has been raised by Haldane (1957). The spread through a population of a gene, initially rare, implies the selective elimination of its allele through the so-called 'genetic death' of the carriers of this newly disfavoured allele. Haldane sought to estimate the number of these 'genetic deaths', which he termed the cost of natural selection. He calculated that cumulatively it would be about 30 times the number of individuals per generation in the population, though this total must obviously be spread over many generations. He concluded therefore that evolutionary changes must be slow, and that not many genes could be spreading at the same time if the population was to survive. These conclusions cannot, however, be accepted uncritically (see Mather, 1969), though it is true that changes in the genetical structure of populations will generally be slow for other reasons that we shall examine in later chapters. Haldane's estimate of the number of genetic deaths as about 30 times the population size springs from the assumptions on which he based his calculations, one of them being that population size is constant. He further ignored the reason for the spreading of a newly favoured gene. This may be because the environment has deteriorated and so has come to disfavour the commonly existing allele, in which case it is not so much the price of natural selection that is being discussed as the price of a change in the environment, which must indeed often lead to extinction of populations. Or it may be because a new and more favourable allele has been introduced by mutation or immigration, in which case the genetic load is being artificially and spuriously raised by the re-definition of fitness in relation to this newly introduced genotype, even though the 'unfit' genotypes (as they are now regarded) are perfectly capable of surviving and reproducing without any risk whatsoever of the population dying out through unfitness. Finally he ignores the conditional nature of most selective forces, and their relation to population size, which as we have seen he unjustifiably assumed to be constant.

A population whose member individuals can average one offspring each under the prevailing conditions can maintain itself irrespective of whether some rare or non-existent genotype could average more. Only if this, or similar genotypes, were to spread could the existing members come to be at a disadvantage; and even then, provided they maintained their average of one offspring each, they would not diminish in absolute numbers, but would constitute a lower proportion of a population which was growing in numbers. When the size of the population had become such that it was limited by external conditions, such as resources of a nutrient, competition would intensify and the earlier types, adequate by themselves, would lose fitness, with the result that their rates of reproduction would fall below maintenance level because of the competition with the later more efficient invaders. They would then tend to die out not because they were inadequate of themselves, but because they were in competition with types which were fitter under the circumstances. Competitive selection is, as we have already observed, a potentially self-regulating force in that it diminishes in intensity if for any reason the population dwindles relative to its resources, and intensifies if the population grows. Only disabilities unrelated to competition will lead to unconditional losses of fitness, and though these exist they must be in a minority. Fitness is basically a comparative quality, relating to the environment in which the individual finds itself, which environment includes the other individuals of the population. And if fitness is comparative, the so-called load cannot be absolute or of unconditional significance. Indeed it is far from clear that the concept, as it has been defined and used, has much significance at all for the understanding of populations and their structure.

4 Theory of Variability

Free and potential variability

Consider two homozygous lines, one of which carries a gene G and the other its allele g: the lines will display the effect of this gene substitution as a difference between their phenotypes. The two lines are then crossed and replaced by their F_1, all the members of which are alike. All phenotypic variation stemming from this gene difference has vanished; but if an F_2 is raised it appears once more with the reappearance of the homozygotes. Clearly, although no variation was detectable, phenotypically in F_1, the capacity for it was latent in the heterozygotes. These may thus be said to carry hidden or potential variability as distinct from the patent or free variability expressed as the phenotypic difference between the homozygous parents.

These changes can be quantified. In measuring the phenotype, let us take as the origin the point mid-way between the two homozygotes in the expression of the character that the gene affects, and to make the calculation simpler let us assume dominance to be absent and the heterozygote's expression of the character therefore to be midway between those of the two homozygotes. Let the GG homozygotes exceed the mid-point, or mid-parent as it is often called, by d in its expression of the character. Then gg will deviate from the mid-parent by $-d$ and Gg will fall on it. So half the parental homozygotes deviated by d and the other half by $-d$, and their variance as a group was thus

$$\tfrac{1}{2}d^2 + \tfrac{1}{2}(-d)^2 = d^2$$

The F_1 consisted solely of heterozygotes with an expression of the character falling on the mid-parent. Their variance round this point was thus 0, which is another way of saying that the phenotypic variation had vanished. The F_2 includes $\frac{1}{4}$GG with expression d; $\frac{1}{2}$Gg with expression 0; $\frac{1}{4}$gg with expression $-d$; and its variance therefore becomes $\frac{1}{4}(d)^2 + \frac{1}{2}(0)^2 + \frac{1}{4}(-d)^2 = \frac{1}{2}d^2$. If we proceed to F_3, which consists of $\frac{3}{8}$GG; $\frac{1}{4}$Gg; $\frac{3}{8}$gg; the phenotypic variance increases to $\frac{3}{4}d^2$. In F_4 it becomes $\frac{7}{8}d^2$ and if the inbreeding is continued until full homozygosis is attained once again the phenotypic variance is restored to its original value of 1 (Table 10). Thus the whole of the

Table 10 The redistribution of variability from Crossing followed by Selfing.

Genotype	GG	Gg	gg	Phenotypic var.	Potential
Phenotype	d	0	$-d$	= Free Variability	Variability
Frequency among					
Parents	$\frac{1}{2}$	0	$\frac{1}{2}$	d^2	0
F_1	0	1	0	0	d^2
F_2	$\frac{1}{4}$	$\frac{1}{2}$	$\frac{1}{4}$	$\frac{1}{2}d^2$	$\frac{1}{2}d^2$
F_3	$\frac{3}{8}$	$\frac{1}{4}$	$\frac{3}{8}$	$\frac{3}{4}d^2$	$\frac{1}{4}d^2$
F_4	$\frac{7}{16}$	$\frac{1}{8}$	$\frac{7}{16}$	$\frac{7}{8}d^2$	$\frac{1}{8}d^2$
$F\infty$	$\frac{1}{2}$	0	$\frac{1}{2}$	d^2	0

variability is not expressed freely in the phenotypic differences until the population has once again come to consist solely of homozygotes. So long as there are heterozygotes some of the variability remains in the potential state, the proportion so hidden being in fact the same as the proportion of heterozygotes in the population.

We are thus recognizing two states of variability, the free variability that is patently expressed in the phenotype and in this simple case corresponds in quantity to the proportion of homozygotes in the population, and the potential variability that does not express itself in the phenotype and in this simple case corresponds in quantity to the proportion of heterozygotes in the population. The free variability of the parent homozygotes was converted into the potential variability of the F_1 by crossing, and this potential variability was reconverted into free by segregation during the generations of inbreeding. These relations may be expressed diagrammatically as in Fig. 4.

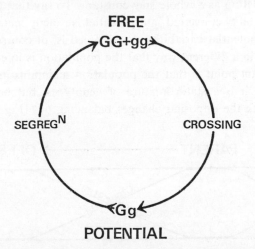

VARIABILITY

Fig. 4 Free and potential variability. The free variability displayed by the differences in phenotype between the two homozygotes, GG and gg, is converted into the potential variability of the heterozygotes, Gg, by crossing, and the potential variability is converted into free by segregation.

In the example we have been discussing, the acts of crossing and segregation were separated in time, crossing taking place to produce the F_1 and segregation proceeding during the production of the later generations. It was in fact this separation that enabled us to distinguish the consequences of each. The two processes, however, can obviously go on simultaneously as, for example, in a random breeding population. Such a population with $u = v = \frac{1}{2}$ has the composition of an F_2 and this continues generation after generation. Yet its heterozygotes are continually segregating, and by giving half homozygotes and half heterozygotes in their progeny are freeing half the potential variability that they contain while retaining the other half in the potential state. At the same time, however, the homozygotes of the population are crossing, not only with like, but also with unlike homozygotes and with heterozygotes, and by giving progeny half of which is heterozygous are converting half the free variability they carry into potential variability. These two changes exactly match in quantity, so the proportions of free and potential variability in the population as a whole stay constant. To say that the amount of free variability converted into potential is exactly matched by the amount of potential variability that is freed is, of course, no more than saying in a different way that the population is in equilibrium. The important point is that the population maintains its structure, not because it is a static mixture of genotypes, but because it is dynamic with the opposing changes balancing out (Fig. 5). In this

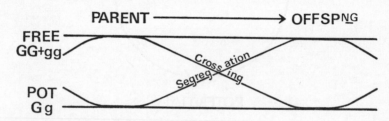

Fig. 5 Variability in a randomly breeding population. In each generation some of the free variability is converted into potential by crossing but at equilibrium this is balanced by the freeing, through segregation, of an equal amount of potential variability. Thus although the genetical constitution of the population is constant there is a flow of variability and the equilibrium is dynamic.

it may be contrasted with a population containing equal numbers of G and g genes, but reproducing by self-fertilization and therefore consisting at equilibrium of equal numbers of the two homozygotes, GG and gg. Such a population retains its structure because it is static and rigid, not because it is dynamic and balanced. The significance of this distinction will become clearer when we have examined a situation involving more than one gene.

Utilized and fixed variability

Before we move on to the more complex case, there is one further aspect of variability to be examined in the simpler situation of the single gene difference. Let us suppose that selection is applied to our random mating population by the elimination of the gg homozygotes as parents in each generation. After such selection in the original population, we are left with GG and Gg in the ratio $\frac{1}{4}:\frac{1}{2}$, which gives a gene frequency of $u = \frac{2}{3}$ and $v = \frac{1}{3}$. The next generation will, therefore, consist of $\frac{4}{9}$GG; $\frac{4}{9}$Gg; $\frac{1}{9}$gg and the mean expression of the character will be

$$x = \frac{4}{9}d + \frac{1}{9}(-d) = \frac{1}{3}d$$

The variance of the character, measuring the free variability, will be calculated round this mean and will thus turn out as

$$\frac{4}{9}d^2 + \frac{4}{9}0^2 + \frac{1}{9}(-d)^2 - \left(\frac{1}{3}d\right)^2 = \frac{4}{9}d^2$$

the correction for the mean being the square of the mean itself because with the frequencies of the classes summing to 1 the sum of squares is also the variance. The proportion of heterozygotes in the population is $\frac{4}{9}$ and consequently the potential variability is $\frac{4}{9}d^2$.

We have thus accounted for only

$$\frac{4}{9}d^2 + \frac{4}{9}d^2 = \frac{8}{9}d^2$$

in terms of free and potential variability. The remaining $\frac{1}{9}d^2$ is represented by the shift which selection has produced in the mean, from 0 to $\frac{1}{3}d$. This represents variability which has been utilized by selection and can, of course, be measured by

$$x^2 = \left(\frac{1}{3}d\right)^2 = \frac{1}{9}d^2$$

thus completing the balance sheet of variability (Table 11).

Table 11 The changing balance sheet of variability under random mating starting with $u = v$, and selectively eliminating all gg individuals in each generation.

Generation	u	v	Frequencies of GG Gg gg with Phenotypes			Mean	Variability		
			d	0	$-d$		Free	Potential	Utilized
0	$\frac{1}{2}$	$\frac{1}{2}$	$\frac{1}{4}$	$\frac{1}{2}$	$\frac{1}{4}$	0	$\frac{1}{2}d^2$	$\frac{1}{2}d^2$	0
1	$\frac{2}{3}$	$\frac{1}{3}$	$\frac{4}{9}$	$\frac{4}{9}$	$\frac{1}{9}$	$\frac{1}{3}d$	$\frac{4}{9}d^2$	$\frac{4}{9}d^2$	$\frac{1}{9}d^2$
2	$\frac{3}{4}$	$\frac{1}{4}$	$\frac{9}{16}$	$\frac{6}{16}$	$\frac{1}{16}$	$\frac{1}{2}d$	$\frac{3}{8}d^2$	$\frac{3}{8}d^2$	$\frac{1}{4}d^2$
3	$\frac{4}{5}$	$\frac{1}{5}$	$\frac{16}{25}$	$\frac{8}{25}$	$\frac{1}{25}$	$\frac{3}{5}d$	$\frac{8}{25}d^2$	$\frac{8}{25}d^2$	$\frac{9}{25}d^2$
⋮	⋮	⋮	⋮	⋮	⋮	⋮	⋮	⋮	⋮
∞	1	0	1	0	0	d	0	0	d^2

If the selection is continued, only the $\frac{4}{9}$GG and $\frac{4}{9}$Gg of the population are left as parents with the result that $u = \frac{3}{4}$ and $v = \frac{1}{4}$. Then the second selected generation comprises $\frac{9}{16}$GG; $\frac{6}{16}$Gg; $\frac{1}{16}$gg, giving a mean of

$$\frac{9}{16}d + \frac{1}{16}(-d) = \tfrac{1}{2}d$$

and free variability of

$$\frac{9}{16}d^2 + \frac{6}{16}0^2 + \frac{1}{16}(-d)^2 - (\tfrac{1}{2}d)^2 = \frac{3}{8}d^2$$

The potential variability of the heterozygotes is also $\frac{3}{8}d^2$ and the tally is completed by the variability utilized in the selective shift of the mean, viz. $\left(\frac{1}{2}d\right)^2 = \frac{1}{4}d^2$. The variability is thus fully accounted for since free + potential + selected $= \frac{3}{8}d^2 + \frac{3}{8}d^2 + \frac{1}{4}d^2 = d^2$. This and the third selected generation are also shown in Table 11.

If the selection is continued indefinitely, gene g will vanish from the population, which will consist entirely of GG and hence will have a mean expression of d. There will be no free variability left and no potential variability either, since there are no heterozygotes. All the variability will have been utilized by selection in shifting the mean from 0 to *d*. Indeed, now that g has vanished we may say that the variability has been fixed by selection, for mutation apart, the population will continue to be wholly GG no matter what selection should be applied, even in the reverse direction. So long, however, as a single g gene remained the variability could not have been said to have been fixed, for reversion by the population to its original composition would have been possible if selection had reversed. Table 11 shows how the free and potential variability decline in step with one another and the utilized fraction of the variability rises as selection operates over the generations until the final state of fixation is reached.

We have now recognized three states of variability, free, potential and utilized, which must among them account for all the variability. Variability of one kind may be transformed into another, potential into free by segregation, free into potential by crossing and free into utilized by selection. Potential variability cannot be utilized directly by selection; it must first have been freed by segregation. Utilized variability may be fixed by the complete elimination of one allele; but until this happens it is reconvertible by a change in selection.

Two types of potential variability

When we turn to consider the more complex case of variation at two loci affecting the same character in the same way, we see that there is still a fourth state of variability to be recognized. Again for simplicity of exposition let us assume that there are a pair of alleles at each of the two loci G – g and H – h, that dominance is absent from both of them and that like alleles at the two loci have the same effect as one another on the character, that is that GG and HH each add *d* to the expression and gg and hh each reduce the expression by *d*, or representing this algebraically add $-d$ to it. If the gene frequencies are all equal, that is $u_g = v_g = u_h = v_h = \frac{1}{2}$ and the genes are unlinked, both an F_2 and a random breeding population will have the composition shown in Fig. 6, which also displays the phenotypes

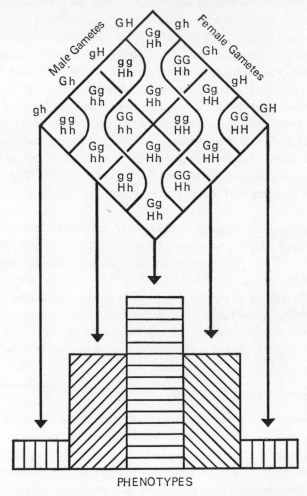

PHENOTYPES

Fig. 6 Polygenic variation. The distribution of phenotypes obtained in an
F₂ (or a randomly breeding population with equal gene frequencies) with
two genes of equal and additive effect but without dominance, neglecting
non-heritable variation. The phenotypic expression is proportional to the
number of capital letters in the genotype. There would be seven such
phenotypes classes with three genes, nine classes with four genes and $2n+1$
classes with *n* genes. (Redrawn with Professor J. L. Jinks' agreement from
K. Mather and J. L. Jinks (1971) *Biometrical Genetics* (2nd edn.) Chapman
and Hall, London).

associated with the nine genotypes and shows how the more central expressions of the character are the commoner, just as we see in continuous variation, for indeed the genes we have postulated constitute the simplest form of polygenic system.

The nine genotypes fall into five phenotypic classes. In GGHH the genes at the two loci are acting in the same direction, each adding an increment d to the expression of the character which thus exceeds the mid-parent by $2d$. The genes at the two loci are reinforcing one another in gghh also, but now the increment added by each is $-d$, making $-2d$ in all. In GGHh, the GG genes are adding d but H and h balance one another out and so add nothing. This genotype, therefore, gives rise to a phenotype of d, as does GgHH for the same reason. Similarly ggHh and Gghh give $-d$. The double heterozygote has a phenotype of 0; i.e. falls on the mid-parent, because G is balanced by g and H by h. It will be observed, however, that the homozygous genotypes ggHH and GGhh also give phenotypes of 0. Here the balancing is not of one allele by the other at each locus as in the double heterozygote; but comes about because in the case of ggHH the increment d prospectively added by HH is cancelled out by the increment of $-d$ prospectively added by gg, and *vice versa* in GGhh.

If we start with the two homozygotes GGHH and gghh as parents the whole of the variation between them is free in the phenotype, the one deviating from the mid-parent by $2d$ and the other by $-2d$. The variability is thus $\frac{1}{2}(2d)^2 + \frac{1}{2}(-2d)^2 = 4d^2$. Crossing these two homozygotes gives the doubly heterozygous F_1, GgHh with a phenotype lying on the mid-parent and no free variability: the variability is all potential and associated with the balancing action of alleles when present together in the heterozygote. The F_2 is of the composition shown in Fig. 6, with $\frac{1}{16}$ GGHH having an expression of $2d$; $\frac{2}{16}$ each of GgHH and GGHh with expression d; $\frac{4}{16}$ of GgHh and $\frac{1}{16}$ each of GGhh and ggHH all with an expression 0; $\frac{2}{16}$ each of ggHh and Gghh with expression $-d$; and $\frac{1}{16}$ of gghh with expression $-2d$. The mean of the F_2 is 0 and its free variability, measured by its variance, is thus

$$\frac{1}{16}(2d)^2 + \frac{4}{16}(d)^2 + \frac{6}{16}0^2 + \frac{4}{16}(-d)^2 + \frac{1}{16}(-2d)^2 = d^2$$

or one quarter of the variability shown by the original parents. Now half the variability, that is $2d^2$, must be present in the potential state characteristic of heterozygotes since half the individuals are heterozygotes in respect of each locus with the consequence that the prospective effects of half the genes are being balanced out by their alleles. We have thus accounted for $d^2 + 2d^2$ or three-quarters of the variability with which we started ($4d^2$) in terms of the states of variability with which we are familiar, viz. free and the potential form associated with heterozygosity. Utilized variability does not come into the account because no selection has been applied and the mean of the F_2 is 0. The remaining quarter of the variability is in fact of a new kind: it is potential variability but of a kind which depends on the balancing action of genes at different loci, and occurs in the homozygotes GGhh and ggHH.

The significance of this new state of variability can be seen even more clearly if the initial homozygous parents are not GGHH and gghh, but ggHH and GGhh. Now the two parents are phenotypically alike, both showing the mid-parental value. On crossing they give GgHh in F_1, and this has a phenotype like those of its parents. Thus in neither parental or F_1 generation is there any free variability. Yet on raising an F_2 free variability appears, and indeed the F_2 is of the same composition with, therefore, the same distribution of variability among the three states (free, potential of heterozygotes and potential of homozygotes) as the F_2 from GGHH × gghh. If we continue to raise F_3, F_4 and so on until complete homozygosis is attained, the population will consist of the four homozygotes GGHH, GGhh, ggHH, gghh, in equal numbers. The free variability of such a population is clearly

$$\frac{1}{4}(2d)^2 + \frac{1}{4}0^2 + \frac{1}{4}0^2 + \frac{1}{4}(-2d)^2 = 2d^2$$

Table 12 Balance sheet of variability for two gene pairs, unlinked, of equal action and without dominance, in a cross followed by selfing without selection.

The initial cross may be either 1. GGHH × gghh or 2. GGhh × ggHH, the later generations being the same for both cases. Further information in the text.

Genotype	GGHH	GgHH	GGHh	GGhh	ggHH	GgHh	Gghh	ggHh	gghh	Variability		
											Potential	
Phenotype	$2d$	d	d	0	0	0	$-d$	$-d$	$-2d$	Free	Het.	Hom.
Parents 1.	$\frac{1}{2}$	—	—	—	—	—	—	—	$\frac{1}{2}$	$4d^2$	0	0
2.	—	—	—	$\frac{1}{2}$	$\frac{1}{2}$	—	—	—	—	0	0	$4d^2$
F_1	—	—	—	—	—	1	—	—	—	0	$4d^2$	0
F_2	$\frac{1}{16}$	$\frac{2}{16}$	$\frac{2}{16}$	$\frac{1}{16}$	$\frac{1}{16}$	$\frac{4}{16}$	$\frac{2}{16}$	$\frac{2}{16}$	$\frac{1}{16}$	d^2	$2d^2$	d^2
F_3	$\frac{9}{64}$	$\frac{6}{64}$	$\frac{6}{64}$	$\frac{9}{64}$	$\frac{9}{64}$	$\frac{4}{64}$	$\frac{6}{64}$	$\frac{6}{64}$	$\frac{9}{64}$	$\frac{3}{2}d^2$	d^2	$\frac{3}{2}d^2$
F_4	$\frac{49}{256}$	$\frac{14}{256}$	$\frac{14}{256}$	$\frac{49}{256}$	$\frac{49}{256}$	$\frac{4}{256}$	$\frac{14}{256}$	$\frac{14}{256}$	$\frac{49}{256}$	$\frac{7}{4}d^2$	$\frac{1}{2}d^2$	$\frac{7}{4}d^2$
.....									.....	.....	.....	.....
F_∞	$\frac{1}{4}$	—	—	$\frac{1}{4}$	$\frac{1}{4}$	—	—	—	$\frac{1}{4}$	$2d^2$	0	$2d^2$

and represents half the variability present in the population. The other half is homozygotic potential, locked up in the genetical difference between GGhh and ggHH which are phenotypically alike. (Table 12).

It will be seen from Table 12 that as inbreeding proceeded the amount of homozygotic potential variability rose in parallel with the free. It therefore sprang like the free variability from the conversion by segregation of heterozygotic potential, which was the only type present in F_1. Similarly the homozygotic potential of parents GGhh and ggHH was converted into heterozygotic potential of the F_1 by crossing. Thus the homozygotic potential state of variability stands in the same relation to heterozygotic as does the free: both are converted into the heterozygotic state by crossing and released from it by segregation. Furthermore, since the F_2 contains both free and homozygotic potential variability no matter whether the original cross was GGHH × gghh or GGhh × ggHH, both are produced side by side from the heterozygotic potential no matter whether this last arose by the conversion of free variability or of homozygotic potential through crossing. These relations among the states of variability are shown diagrammatically in Fig. 7.

This diagram illustrates a further point also. Though homozygotic potential is like free variability in its relation to heterozygotic potential, it is like heterozygotic potential itself in another crucial respect: it is variability that is hidden in the genotype and therefore inaccessible to the direct action of selection. In our simple model GGhh and ggHH have the same phenotypes and they cannot, therefore, be acted on differentially by selection any more than can the F_1 consisting of GgHh. Only in the F_2 does free variability appear and selection become effective. All potential variability must be converted to the free state before selection can utilize it, and this takes longer with homozygotic than with heterozygotic potential since crossing is the necessary precursor of the segregation that occurs only from heterozygotes. We can see now the chief reason why the dynamic quality of the equilibrium in an outbreeding, as opposed to inbreeding, population is important. Only by crossing and the subsequent segregation can a major part of the potential

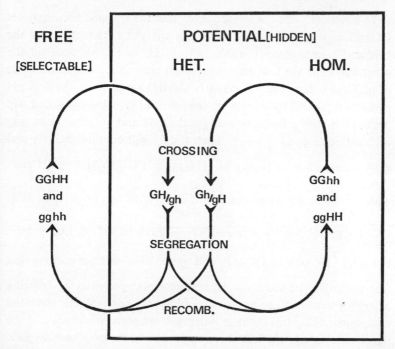

Fig. 7 The states of variability. Free variability is expressed as phenotypic differences among the genetic classes, heterozygotic potential variability is concealed by the balancing effects of alleles and is therefore a property of heterozygotes, and homozygotic potential is concealed by the balancing effects of non-allelic genes. Free variability is available to the action of selection, but both types of potential variability (enclosed in the box) are incapable of being acted on by selection. The arrows show the directions of flow of the variability, the rates of flow being governed by crossing, segregation and recombination.

variability be freed and made available to selection. An inbred population consisting of all possible types of homozygote contains nearly as much potential variability as an outbreeding one carrying the same genes with the same frequencies; but it is a static, rigid population since this potential variability is all homozygotic and all incapable of being freed.

One further important point must be made about this flow of variability from one state to another. We have assumed G-g and H-h

F

to be unlinked, with the consequence that the extreme homozygotes GGHH and gghh occurred with just the same frequencies as the balanced homozygotes, GGhh and ggHH, in F_2. Suppose on the other hand that the two genes are linked with a recombination value of p. Then if the parental cross is GGHH × gghh, the Gh and gH gametes produced by F_1 will be the result of crossing-over, and will occur each with a frequency of $\frac{1}{2}p$, the GH and gh gametes having each a frequency of $\frac{1}{2}(1-p)$. Then the F_2 will contain GGHH and gghh each with a frequency of $\frac{1}{4}(1-p)^2$ but GGhh and ggHH each with the lesser frequency of $\frac{1}{4}p^2$. Similarly, if the original cross is GGhh × ggHH the F_2 carries $\frac{1}{4}(1-p)^2$ each of GGhh and ggHH, but only $\frac{1}{4}p^2$ each of GGHH and gghh. Thus linkage ensures that free variability turned into the heterozygotic potential by crossing emerges preponderantly as free variability, the homozygotic potential component being less because it depends on recombination, and the tighter the linkage the less this component is. Equally homozygotic potential variability entering the heterozygotic state is freed only by recombination. It thus tends to emerge preponderantly as homozygotic potential, the free component diminishing as the linkage tightens. In our simple example the actual proportions of free variability and homozygotic potential in an F_2 are in fact:

Initial Cross	Free	Hom. Pot.
GGHH × gghh	$2(1-p)d^2$	$2pd^2$
GGhh × ggHH	$2pd^2$	$2(1-p)d^2$

showing direct dependence on the recombination fraction in their relative magnitudes.

In a random mating population under no selection and at equilibrium the four types of gamete are produced with frequencies depending only on the gene frequencies and independent of the recombination value. Where the frequencies of G and g are u_g and v_g and of H and h are u_h and v_h, the gametes occur in the proportions

$u_g u_h$ GH; $u_g v_h$ Gh; $v_g u_h$ gH; $v_g v_h$ gh. The two double hetero-zygotes, coupling GH/gh and repulsion Gh/gH, will therefore both occur with frequency $2u_g v_g u_h v_h$ in the population. The recombinant gametes from the one type of double heterozygote are the parental combinations from the other and their excesses and shortfalls thus balance, with the consequence that the overall gametic output is again independent of the recombination value. Thus the distribution of variability between free and homozygotic potential is independent of linkage relations in such a population at equilibrium. But if selection is acting to diminish the frequencies of the extreme combinations of genes and hence of coupling double heterozygotes, there will be an overall flow of variability from the homozygotic to the free. This flow will not be independent of linkage relations: it is at its highest when $p = 0 \cdot 5$ and recombination is free, and it diminishes with p until it becomes very small indeed with tight linkage. In other words, the tighter the linkage the less readily is homozygotic potential variability freed and the longer the time necessary for it to become available for utilization by selection. Thus recombination, like crossing and segregation, is important in controlling the flow of polygenic variability among the different states in which it can appear (Fig. 7).

Unequal gene frequencies

The model we have been using to bring out the properties of variability is a very simple one and it requires elaboration in several ways if it is to be generalized. In the first place we have assumed equal gene frequencies at both loci, that is $u_g = v_g = u_h = v_h$. If $u_g = u_h \neq v_g = v_h$ the mean of the population will no longer be 0 the mid-parent value which we have taken as the origin. Instead it will be $2d(u-v)$, which of course becomes 0 when $u = v$. There is thus an element of bias, or, as we have called it when discussing the effects of selection, utilized variation, and this will amount to $4d^2(u-v)^2$. Since total variation in the system is as we saw above $4d^2$, this leaves $4d^2[1-(u-v)^2] = 16d^2uv$ to be accounted for by the other states of variability. Now there will be $2uv$ heterozygotes in respect of each locus in the population and as the alleles cancel out one another's effects in a heterozygote, $2uv$ of the total variation, i.e. $2uv \times 4d^2 = 8uvd^2$, will be in the heterozygous potential state. It is easy to show that the free variation, measured as the variance round the population mean of $2d(u-v)$, is $4uvd^2$ and this leaves $(16-8-4)uvd^2 = 4uvd^2$ in the homozygotic potential state, and we see that variability appears in the free, heterozygote potential and homozygote potential states in relative amounts $\frac{1}{4} : \frac{1}{2} : \frac{1}{4}$ just as when $u = v$, the difference being that these three states now account for $16uvd^2$ of the total $4d^2$ variability measured round the mid-parent of 0, the remaining $4d^2(1-4uv) = 4d^2(u-v)^2$ being taken up by bias of the mean or the utilized variability according to how we look at it. When $u = v = \frac{1}{2}$ is substituted these values reduce to those found earlier.

The situation becomes rather more complicated when $u_g \neq u_h$. We have defined the total variability by considering a population consisting solely of the extreme homozygotes, GGHH and gghh, and finding its variance round the mid-parent. Since all individuals in such a population depart from the mid-parent by either $2d$ or $-2d$, this total variability must be $4d^2$, of which as we have seen $4d^2(u-v)^2$ is taken up by the departure of the mean from the mid-parent, the rest being divided up in the proportion $\frac{1}{4} : \frac{1}{2} : \frac{1}{4}$ between the other states of variability. When $u_g \neq u_h$ however a fourth factor enters in because the population could not consist entirely of extreme homozygotes. Let $u_g = u + e$ and $u_h = u - e$. Then the maximum number of extreme homozygotes possible will be $u - e$ of GGHH and $v - e$ of gghh, leaving $2e$ to be either GGhh or ggHH both of which fall on the mid-parent. The mean is thus lower by $2d.2e$ than it would have been if $u_g = u_h = u$. Thus not only is $4d^2(u-v)^2$ taken up by the departure of the mean due to any difference between u and v, but an additional $16d^2 e^2$ by the difference between u_g and u_h, the gene frequencies at the two loci. This leaves $16d^2(uv-e^2)$ to be accounted for by the free, heterozygotic potential and homozygotic potential states, and it is not difficult, even if algebraically a little tedious, to show that once again this remaining variability is divided between these three states in the proportions $\frac{1}{4}:\frac{1}{2}:\frac{1}{4}$. The proportions are thus a characteristic of the two gene case (Table 13).

When, however, we move to a system with more than two genes the relative proportions of the free and homozygotic potential states change. Taking the case of three genes of like effect and assuming, for simplicity, equal gene frequencies, the two extreme homozygotes

Table 13 Balance sheet of variability in a randomly breeding population, for two gene pairs of equal effect without dominance. Further information in the text.

State of Variability	Amount	Proportion	Special cases $u_g = u_h$	$u_g = u_h = \frac{1}{2}$
Total	$4d^2$	1	1	1
Bias due to $u \neq v$	$4d^2(u-v)^2$	$(u-v)^2$	$(u-v)^2$	0
Bias due to $u_g \neq u_h$	$16d^2e^2$	$4e^2$	0	0
Free	$4d^2(uv-e^2)$	$(uv-e^2)$	uv	$\frac{1}{4}$
Het. Potential	$8d^2(uv-e^2)$	$2(uv-e^2)$	$2uv$	$\frac{1}{2}$
Hom. Potential	$4d^2(uv-e^2)$	$(uv-e^2)$	uv	$\frac{1}{4}$

will depart from the mid-parent by $3d$ and $-3d$ respectively, giving a total variability of $9d^2$. A half of this will be heterozygotic potential as before, since in respect of the third locus, as of the other two, half the individuals will be heterozygotes. This leaves $\frac{9}{2}d^2$ to be accounted for. A simple calculation reveals that the free variability amounts to $\frac{3}{2}d^2$ and the homozygotic potential must then be $3d^2$, or twice the free variability in amount. This calculation can be generalized. With n genes all of effect d, the extreme homozygotes depart by $\pm nd$ from the mid-parent and the total variability is n^2d^2, of which half will be heterozygotic potential leaving $\frac{1}{2}n^2d^2$ to be accounted for. The free variability accounted for by these genes is $\frac{1}{2}S(d^2) = \frac{1}{2}nd^2$ (see Mather and Jinks, 1971), and this leaves $\frac{1}{2}n(n-1)d^2$ for the homozygotic potential. Thus the ratio of free to homozygotic potential is $\frac{1}{2}n : \frac{1}{2}n(n-1)$ or $1 : n-1$. The hidden variability clearly rises with the number of genes in the system (Fig. 8).

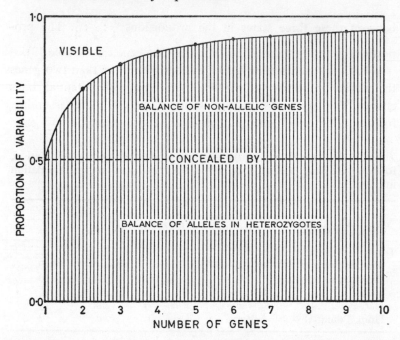

Fig. 8 The proportions of variability which are freely expressed and concealed in the genotype in relation to the number of genes in the system. The genes are assumed to be of equal effect and with equal frequencies in the population. Variability concealed by the balance of alleles is heterozygotic potential and is at a constant proportion of 0·5. Variability concealed by the balance of non-alleles is homozygotic potential and rises with the number of genes in the system, just as the proportion of visible or free variability falls. (Reproduced with the permission of the publishers from K. Mather (1964) *Human Diversity*, Oliver and Boyd, Edinburgh).

In the case of two genes with equal effects, the ratio 1 : 2 : 1 for free: heterozygotic potential: homozygotic potential variability applies whatever the gene frequencies may be at the two loci (Table 13). This suggests that with genes of equal effect the ratio $1 : n : n-1$ we have just found will also apply whatever the gene frequencies may be.

Dominance

We have been assuming for simplicity of exposition that all the

genes were of equal effect. Where they are not, the extreme homozygotes will depart from the mid-parent by $S(d) = nd$, where d is
the mean of d_g, d_h, etc.; and the total variability will be n^2d^2. Half
of this will of course be heterozygotic potential, leaving $\frac{1}{2}n^2d^2$ for
the combined free and homozygotic potential variability. If now
we let

$$d_g = \bar{d}(1+\alpha_g); d_h = \bar{d}(1+\alpha_h) \text{ etc.}$$

it can be shown that the free variability amounts to $\frac{1}{2}S(d^2) =
\frac{1}{2}n\bar{d}^2(1+V_\alpha)$ where V_α is the variance of α (Mather, 1949; Cooke
and Mather, 1962; Mather and Jinks, 1971). This leaves

$$\frac{1}{2}n^2\bar{d}^2 - \frac{1}{2}n\bar{d}^2(1+V_\alpha) = \frac{1}{2}n(n-1-V_\alpha)\bar{d}^2$$

for the homozygotic potential, which is thus $(n-1-V_\alpha)/(1+V_\alpha)$
times the free variability in amount. Inequality of the effects of the
genes thus increases the proportion of variability that is expressed
freely as phenotypic differences, as would indeed be expected since
inequality precludes the exact cancelling out among the genes of
one another's effects on the phenotype; but unless the number of
genes in the system is very small, the homozygotic potential will still
greatly exceed the free variability in quantity.

Another of our simplifying assumptions has been that dominance
was absent, the phenotype of each heterozygote falling exactly
mid-way between those of the corresponding homozygotes with the
multiple heterozygote characteristic of F_1 lying on the mid-parent
value. Obviously this need not, and indeed generally will not be the
case. Let the heterozygote for any gene depart by h from the mid-
value of the two homozygotes, where h can take sign according to
whether the allele making for greater expression of the character,
or that making for the lesser expression is dominant. Then the
mean of F_1 will depart from the mid-parent by $S(h)$, the mean of F_2
(or the mean of a randomly breeding population with equal gene
frequencies) will depart by $\frac{1}{2}S(h)$, the mean of F_3 by $\frac{1}{4}S(h)$ and so on.

The h's also add to the free variability, their contribution being
$\frac{1}{4}S(h^2)_2$ in F_2 or in a random breeding population with equal gene

frequencies; but this contribution is quite independent of and additional to the free variability springing from the *d* effects of the genes (see Mather and Jinks, 1971). This variability springing from the *h*'s – the *H* variability as Mather and Jinks call it – has, however, a significance different from the *D* variability, springing from the *d*'s; for whereas the *D* variability is fixable, in that under selection it can bring about permanent changes uniformly affecting all members of the population when these have all become homozygous for one or other allele at a locus, the *H* variability is by definition unfixable in this sense. Thus in considering evolutionary changes we must be concerned primarily with *D* variability and only in perhaps a few special cases with *H*. The variability whose states and partitioning we have hitherto been discussing in this chapter is of the *D* kind and, although the immediate impact of selection may be modified or vitiated by the *h*'s, in the long run partitioning and utilization of *D* variability are unaffected by them. The long term potentialities and fate of a population depend therefore on the fixable *D* variability and not on the unfixable *H*. Methods exist for the separation of the fixable and unfixable components in estimates of free variability, but they are not always easy to apply, and the estimate especially of the unfixable component may be subject to a large standard error (see Mather and Jinks, 1971). In so far as we cannot estimate them separately, the *H* variability will inflate our estimate of the free variability due to the *d*'s and will, therefore, tend to lead us to under-estimate the hidden variability.

The fixable and unfixable effects of the gene make independent contributions to the variability when the gene frequencies are equal at each locus, i.e. $u_g = v_g = u_h = v_h = \ldots = \frac{1}{2}$; but when this is no longer true for a locus, the *d* and *h* associated with that locus no longer contribute independently to the variation. Their inter-relation in the expression for the amount of free variability in a random breeding population where the gene frequencies are not equal is well known (see for example Mather, 1949; Falconer, 1960; Mather and Jinks, 1971); but the effect on the assessment of the hidden variability has yet to be investigated, though it is hardly likely to alter in any serious way the general conclusions to which we have been led by our present consideration.

Mutation and drift

Before we leave our discussion of variability two further factors must be noticed. We have assumed that the amount of variability is fixed and that variability used up by selection is taken out of the system which is thereby permanently depleted in the sense that the variability available for future change under selection is reduced. This obviously cannot be the whole truth over the long term as it would imply that the amount of variability available for prospective evolutionary adjustment was steadily dwindling and must eventually run out. It also begs the question of where the variability we can detect in populations, originated. Variability must in fact have originated in the past and be continuing to arise in the present by some form of mutation. Mutation is familiar chiefly from the study of genes where changes have brought about readily detectable and virtually always deleterious effects on development, lethal mutation being indeed one of the chief subjects of such studies. Mutation is more difficult to observe in polygenic systems where the individual genes are so very much more difficult to isolate and their effects to observe. The occurrence of such mutation in *Drosophila* has, nevertheless, been established and the increment measured that it adds to the variation in each generation (Clayton and Robertson, 1955; Paxman, 1957). While it must remain true that in the long term all polygenic variability must have arisen in this way, the increments are so small as to be negligible over the short term: Clayton and Robertson's estimate is that in each generation mutation adds somewhere between 1/500 and 1/1000 to the variation in sterno-pleural chaeta number, the character they studied, in the population of *Drosophila* from which they drew their experimental flies. Thus response to selection is no more governed by mutation in the short term than is the availablity of water from a reservoir is by the inflow from the stream that maintains and replenishes it.

A further assumption we have made is that the gene frequencies at each locus remain constant apart from the effects of any selection that might be applied. Now even apart from the effects of selection, the frequencies of the genes will show sampling variation in their transmission from one generation to another. With large populations such sampling variation will be negligible, but with smaller populations this need not be the case; and we should recall that the effective size of the population is the number of parents who contribute offspring to the next generation, rather than the total number of individuals who may appear in it. Wright (1931, 1940 and many other papers) has drawn attention to the prospective importance of changes in gene frequency which result from sampling variation and are hence independent of selection. He has pointed out that such random drift (as he called it) of the gene frequencies could lead to changes that were not favoured, and even opposed, by selection, yet could be the basis for the later development under the impetus of selection of a new genetic structure for the population. It is clearly not possible to set in absolute terms a population size above which random drift would be ineffective, for this would depend on the length of time over which the population was at risk of such change and on the magnitude of any selective forces that might be in operation. It would, however, seem likely that changes of any significance as a result of drift could hardly be expected in populations whose effective size was much over 100. Certain data recorded by Dobzhansky and Wright (1947) make possible a rough estimate of the effective size of populations of *Drosophila pseudoobscura* in California, though it is not easy to define of what a population consists in a species which would appear to have a virtually continuous distribution over large tracts of country.

They released genetically marked flies in the wild and some ten months later found their descendants in the main to be clustered in the general area of the original release. Nevertheless many of the descendants were at greater distances and it was estimated that over half of them must have been half a kilometer or more away. Dobzhansky and Wright also estimated the density of wild flies in this area to be 0·4 per 100 square meters. Thus from the point of view of gene migration the effective size of the population through which a gene could spread in a year must have been at least of the order of the area of a circle of radius 0·5 km multiplied by 0·4 flies for each 100 m^2, or say 10^3 to 10^4 individuals. Despite the obvious uncertainties of such a calculation, this suggests that the effective interbreeding unit is too large for random drift to be of any significance in determining the genetic constitution of the flies over the long term.

A more precisely defined population, of the tiger moth *Panaxia dominula* at Cothill near Oxford, was studied over many years by Fisher and Ford (1947) and Sheppard (1953). This colony of *Panaxia* is polymorphic for a gene which produces the *medio-nigra* phenotype when heterozygous and *bimacula* when homozygous, these two phenotypes being distinguishable from one another as well as from the normal or typical form. A census could thus be made of the gene by inspection in any sample that was caught and the gene frequency in the population thus estimated. At the same time as the gene frequency was being observed, estimates of the size of the population were made by the method of capture, mark, release and recapture. The outcome of these observations over the years 1939 – 1952 is given in Table 14 and calculations made up to 1946 by Fisher and Ford and extended to 1952 by Sheppard showed that the changes in gene frequency are much too large to be accounted for by sampling variation in a population of this size (see also Ford, 1971, for

Table 14 Population size and gene frequency of *medio-nigra* in the Cothill population of *Panaxia dominula*.

(Fisher and Ford, 1947; Sheppard, 1953).

Year	Size of population in thousands	*medio-nigra* gene frequency (%)
1939	—	9·2
1940	—	11·1
1941	2·0 – 2·5	6·8
1942	1·2 – 2·0	5·4
1943	1·0	5·6
1944	5·0 – 6·0	4·5
1945	4·0	6·5
1946	6·0 – 8·0	4·3
1947	5·0 – 7·0	3·7
1948	2·6 – 3·8	3·6
1949	1·4 – 2·0	2·9
1950	3·5 – 4·7	3·7
1951	1·5 – 3·0	2·5
1952	5·0 – 7·0	3·6

detailed discussion). Selective forces must thus be at work on the gene: the precise nature of these forces were not however clear though *medionigra* appeared to be at a disadvantage of 8 % by comparison with the typical form, the precise level of disadvantage probably fluctuating from year to year.

There is indeed little evidence that drift is of any real importance in determining the genetical structure of populations, except perhaps in probably one particular connection. Where a new population of a species or form comes into being in territory previously unoccupied by it, the number of founding immigrants may be small and they may as a result of drift, therefore, depart in their gene frequencies from the population whence they originated. The new population may thus have its genetical constitution determined at any rate in part, by the accidents of sampling in its foundation (Mayr, 1954). The high incidence of the genetically determined disability, porphyria variagata, in certain South African white people has been attributed to this 'founder principle' as it is called and, indeed, the condition can be traced back to a particular member of the early group of settlers in the 17th century (Dean, 1969). Similarly the pattern of genetic variation in the present population of Finland provides evidence of its descent from small locally isolated groups in the past (Nevanlinna, 1972).

Selection experiments

The most striking features of polygenic variability that have emerged from our consideration are the great proportion of it that is hidden in the genotype by combinations in which genes balance one another's effects, and the importance of crossing, segregation and recombination in governing the release of this hidden variation and making it available to the action of selection in bringing about the continuing readjustment of the organism to meet the demands of its environment. We might note, too, that although the heterozygotic potential variability can be a property only of diploid (or polyploid) organisms, this restriction does not apply to the homozygotic potential where the balancing is of genes at different loci. This type of potential variability can obviously exist in haploid forms also, and indeed in chromosomes, such as some concerned with sex-determination, which exist in the haploid condition alongside others in the diploid condition in at any rate certain members of a species.

The properties of the hidden store of variability can be investigated by the application of appropriate selection under experimental conditions and over the last thirty years there is indeed a literature, too extensive to be cited in its entirety, of selection experiments chiefly with *Drosophila melanogaster* which tell us of these properties. From the earliest (Mather, 1941) to the more recent (for example, Thoday and Boam, 1961; Spickett and Thoday, 1966) these experiments agree in showing very large responses in the manifestation of the character for which selection was practised, whether the flies with which the experiment was started came from the wild or from a cross between laboratory stocks. Mather (1941) raised the average number of abdominal chaetae from just over 41 to more than 47 in 8 generations of selection and reduced it to less than 31 by 8 generations of selection in the opposite direction starting from the same F_2. Thoday and Boam (1961) were successful in raising the average number of sternopleural chaetae from 20 to 45. This average well transcends the range of variation that one normally sees in flies caught in the wild or raised in laboratory cultures, and indeed all these experiments give a superficial appearance of selection creating its own variation in producing the great advances that are achieved. Such an appearance is of course exactly as is to be expected where the response to selection is drawing not just on the free variation of the initial flies, but on a larger store of hidden variation which comes available by recombination over the generations. Furthermore, the response to selection is protracted and spread over many generations. Mather (1941) observed an advance for two generations, then a pause of two further generations before an even larger advance began which went on for another four generations. In another selection for increased number of abdominal chaetae, Mather and Harrison (1949) observed the average number of abdominal chaetae to increase steadily from 36 to nearly 56 over twenty generations. Such a protracted response must imply recombination not only of genes in different chromosomes, but also the reassortment of combinations of genes by crossing-over within chromosomes, and indeed by the use of appropriate assaying techniques, Mather and Harrison were able to demonstrate these changes in activity of the individual chromosomes. All of the major chromosomes, X, II and III, showed these internal recombinations and in a later experiment, Breese and Mather (1957) were able to demonstrate that every one of the five regions into which they were able to divide chromosome III, was actively affecting the number of abdominal chaetae and was, therefore, prospectively involved in the internal recombinations upon which depended the major part of the response to selection.

Mutation is, of course, another prospective source of variability; but its inadequacy to account for the great advances obtained when selection is practised in heterogenic material, is demonstrated not only by the observations of Clayton and Robertson and Paxman, to which reference has already been made, but also by the relative ineffectiveness of selection to change chaetae number when it is practised in inbred, homogenic lines (Mather, 1941; Mather and Wigan, 1942). Clearly recombination is the key to the large store of potential variability and hence to response to selection.

As we have seen, the more the genes in the system the greater the store of variability relative to the variation visible as differences among the individual phenotypes. Only by recourse to the special techniques which Thoday (1961) has developed is it possible to locate members of polygenic systems and so count the individual genes, and even these techniques become too demanding to use in tracing genes of relatively small effect. We can thus do no more than arrive at a minimum figure for the number of genes in the system. In the case of the polygenic system governing the normal variation in abdominal chaetae, the evidence from the many experiments on this character would indicate that the number of genes is likely to be nearer 20 than 10 and may well be higher. Indeed, from a single set of experiments Davies (1971) has evidence of genes at 15 loci at least affecting the number of sternopleural chaetae and 14 or 15 affecting the number of abdominals. In mice Falconer (1971) has estimated that some 80 gene differences affecting litter size were present in the population from which he successfully selected for this character.

Spickett and Thoday (1966), however, obtained evidence that the bulk of the effect of the selection for sternopleural chaeta number was traceable to a few locatable units of relative large effect – so large in fact that it is hard to believe that they were segregating in the flies with which the experiment was started, where variation was too small to be compatible with their presence. Where they came from is thus still unclear. It can hardly have been by mutation, not merely for the reasons which we have already discussed, but also because the properties of these units in interaction were so adjusted to the demands of selection as not to be expected from mutation affecting chaetae number in a random way (see Chapter 7). It may well be that they too were the products of recombination between units of lesser effect situated at closely adjacent loci and, therefore, recombining rarely; for a close combination of genes built up by rare recombination would be broken down only by rare recombination and would thus be as persistent in subsequent inheritance as it was difficult to build up in the first place. The nature of such units as Thoday has isolated is eminently worthy of further investigation for they could hold the clue to the evolution of new genes.

5 Types of Selection

Stabilizing selection

The effects of selection on the distribution of genes in a population will depend on the extent to which the individual phenotypes associated with the various genotypes meet or fail to meet the demands of the environment. If, therefore, we are to understand the interplay of selection and variation on which the genetical structure of a population depends, we must examine the various ways in which selection can impinge on a range of phenotypes and, in particular, on a range showing continuous variation in respect of the character or characters in question.

When a character is varying in its expression from one individual to another, some level of its expression must be more consonant than others with the requirements of the environment, and as a consequence the individuals displaying the character to such a level will have a selective advantage over their fellows. We are thus led to the notion of an optimal expression of the character, that is one which, within the extant range of variation, best fits the individual to meet the requirements set by the environment and in so doing puts that individual at a selective advantage. Other individuals will be at a disadvantage relative to those with the optimal expression, and the greater their departure from the optimum the greater the disadvantage.

This concept of the optimal phenotype requires a little elaboration. The demands made by the environment must be expected to vary at any rate to some extent over the range of territory the population is occupying because the micro-environment in one place will not be exactly like that in another. Furthermore, the environment will fluctuate over time in ways that reflect the vagaries of the seasons, and with it the demands it makes will vary also. Thus the environment in which an individual may find itself is neither completely uniform nor completely constant and the optimal phenotype will thus be the one that on the average best fits the individual possessing it in the range of circumstances that it may encounter. There is thus an element of probability, or to put it another way, a statistical quality about the concept of the optimal phenotype, for the environment as well as the phenotype will be varying. It should be emphasized, too, that an optimal phenotype does not imply a single optimal genotype, for it is an essential property of polygenic systems mediating continuous variation that a number, even a sizeable number, of genotypes can give closely similar if not identical ranges of phenotypes. We speak of ranges of phenotypes because non-heritable agencies join with the genotype in determining the phenotype, which is thus not constant even for any single genotype.

Now the optimal phenotype may coincide with, or at any rate closely approximate to, the average phenotype of the population (Fig. 9). Phenotypes departing from the mean will then be at a disadvantage and the greater the departure the greater this disadvantage will be. The primary effect of selection will then be not to change the mean expression of the character, but to reduce the variation it shows, for not only will genotypes which give more extreme expressions be at an immediate disadvantage, but genotypes giving the favoured central phenotypes will in the longer term be the fitter the less their progeny deviate from this mean. In other words individuals giving rise to progeny which not merely reproduce the parental phenotype on the average, but do so with minimal variation among themselves, will leave the greatest number of grandchildren, great-grandchildren, and so on. Thus when the optimum approximates closely to the mean phenotype, selection will tend to stabilize the population at this mean, and it is hence termed stabilizing selection. Such selection will favour the regular reproduction of genotypes giving the fittest phenotypes and so will make for genetical invariance.

G

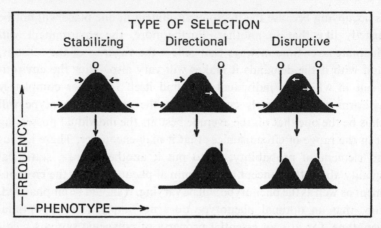

Fig. 9 The three elemental types of selection. O indicates the optimal phenotypes towards which selection is acting, and the direction and force of selection is shown by the arrows in the upper row of phenotypic distributions. The lower row of distributions shows the consequences of the types of selection (reproduced by permission of the Society for Experimental Biology from K. Mather (1953) The genetical structure of populations, *Symp. Soc. Exp. Biol.*, **7**, 66 – 95.)

A number of examples of stabilizing selection at work have been observed. They go back to Bumpus (1899) and Weldon (1901). Bumpus noticed that sparrows suffering from the effects of a storm were the more likely to survive the nearer certain bodily measurements, that he was able to make on them, approximated to the mean of the group that he observed. Weldon's observations were made on snails that he obtained from Plymouth Sound. Because each snail shell constitutes, so to speak, a permanent record of its own development, he was able to compare the variation in the diameter of shell in immature individuals, with the variation shown at the corresponding stage of development as preserved in the shells of individuals that had survived to maturity. The sample that had survived was less variable than the immature snails; evidently the mean expression had been favoured by selection and the more extreme expressions penalized.

One of the best documented examples of stabilizing selection is afforded by Karn and Penrose's (1952) observations on the neo-natal mortality of babies in relation to their birth-weights. In all 13 730 babies, 7037 boys and 6693 girls, born in a London hospital were observed. The distribution of birth-weights is, of course, slightly different for the two sexes, boys being on average slightly heavier than girls. The sexes have, however, been combined for ease of presentation in Fig. 10 which shows the distribution of neo-natal

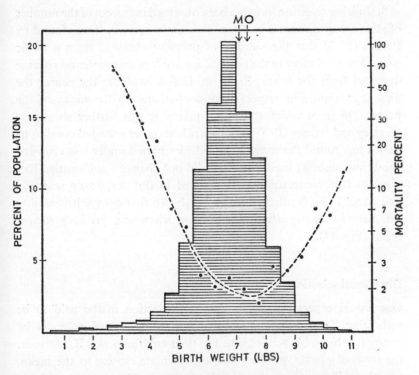

Fig. 10 The distribution of human birth-weights and the rates of neo-natal mortality in the various birth-weight classes. The hatched histogram shows the distribution of birth-weight. The curve shows mortality in relation to birth-weight, the points to which the curve is an approximation showing the values actually observed, class by class. Mortality is set out on a logarithmic scale for ease of presentation. M marks the mean birth-weight and O that with the lowest mortality, and hence *prima facie* the optimum. (The original data are from Karn and Penrose (1952) and the figure is reproduced with the publishers' permission from K. Mather (1964) *Human Diversity*, Oliver and Boyd, Edinburgh).

mortality superimposed on that of birth-weight, the weights being grouped in ranges of ½lb for the purposes of analysis and presentation. The rate of mortality is minimal at a weight approximating closely to the mean, and rises sharply on each side of it. The causes of death no doubt differ between over-large and over-small babies; but that is irrelevant to our present consideration. Whatever the causes of death, stabilizing selection is at work.

Stabilizing selection has also been observed in respect of the number of sternopleural chaetae in *Drosophila melanogaster*. It was found by Barnes (1968) that the number of progeny obtained from a single pair mating fell away as the parents' numbers of sternopleural chaetae departed from the mean. Evidently fitness is higher the nearer the flies approximate in respect of this character to the mean of the population from which they were taken. It was further shown by Kearsey and Barnes (1970) that flies raised under crowded conditions varied less round the mean than did flies varied under less crowded conditions, though the mean itself did not change significantly. This suggests that competition is implicated in this stabilizing selection and McGill and Mather (1971) have shown that competitive ability can vary in *Drosophila* in the way which would produce such a result (Fig. 11).

Directional selection

One would expect the major effect of selection in the wild to be stabilizing, for one would assume that most organisms must be at any rate tolerably well adjusted to their environments. If, however, the optimal phenotype does not approximate closely to the mean, selection will most favour some phenotype, and hence the associated genotypes, other than that which characterizes the mean. The individuals with these deviant but more favoured genotypes will contribute relatively more to the next generation, and other things being equal the mean of the offspring will, therefore, be nearer to the optimum than that of the parents. There will then be a directional element in the selection (Fig. 9). Where (as in the great majority of cases of selection carried out in laboratory experiments or in plant and animal

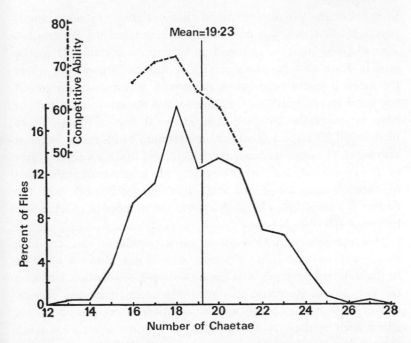

Fig. 11. Sternopleural chaetae and competitive ability in *Drosophila melanogaster*. The solid line shows the percentage of flies with the various number of chaetae among 215 individuals from an F_2 between the Samarkand and Wellington inbred lines. The broken line shows the competitive abilities of the flies with the various number of chaetae, measured by reference to a standard tester line and with the observed proportion subjected to the angular transformation. (Reproduced by permission of the Editor and Publishers of *Heredity* from A. McGill and K. Mather (1971) Competition in *Drosophila*: A case of stabilizing selection, *Heredity*, **27**, 473 – 478).

breeding programmes, though probably much more rarely in nature) the optimum phenotype lies, almost by definition, beyond the range of the character's expression in the material under selection, the selection will be wholly directional in its nature. In so far, however, as the optimum is not so extreme as to lie right outside the range of the character displayed by the population, some element of stabilizing selection will still be present: the forces of selection will be tending to move the mean towards the optimum and simultaneously tending to stabilize the population round that optimum. In other words selection will combine a directional with a stabilizing component. This situation must be common under natural selection. It can be seen in Karn and Penrose's results on human birth-weight where the mean is just a little below the weight which gives the lowest neo-natal mortality (though, of course, this departure may itself be effectively cancelled by some other source of loss of fitness, yet to be detected, in respect of which the optimum birth-weight is below the mean). The same is revealed by McGill and Mather's observations on *Drosophila*, where competitive ability was greatest not at the mean of sternopleural chaetae but at a slightly lower number, with the decline in competitive ability sharper as the number of chaetae rose than as it fell (Fig. 11).

That the selective forces acting on a population can combine directional and stabilizing elements, should not however blind us to the different demands that these two types of selection make on the population. Response to directional selection implies change, and in its turn this change implies variation; a population cannot adjust itself in the way demanded by directional selection without there being free heritable variation present. Directional selection requires, and therefore favours, the flexibility of genetical variation. Stabilizing selection on the other hand will best be met by, and will therefore favour, genetical invariance.

The response of a population to directional selection was examined mathematically in relation to the supreme, capital character, fitness itself, by Fisher (1930), who propounded what he termed the fundamental theorem of natural selection, that 'The rate of increase in fitness of any organism is equal to its genetic variance in fitness at that time' where genetic variance is what we have called D type variability. The theorem applies strictly only where the population is continuous in the sense that the generations overlap, and also strictly only to fitness itself. It applies to other characters only by virtue of the contributions they make to fitness and the relations they show with it. These relations can be complex and we will return to discuss them in a later section. In many selection experiments they are, however, artificially simplified. In such experiments the contri-

bution made by any individual to the next generation will depend
almost entirely on its phenotype in respect of the character for which
the experimenter is practising selection: above a certain level of
expression all individuals are taken as parents, and are given equal
chances of their contributions, whereas below that level all individuals
are rejected and denied any chance of contributing progeny. The
relation of the character in question to fitness is thus in principle at

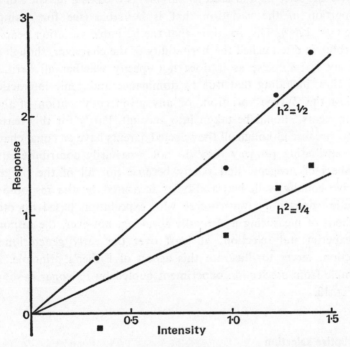

Fig. 12 Response to selection in relation to the intensity of selection for
abdominal chaetae number in *Drosophila melanogaster.* The response to
selection (i.e. departure of mean chaetae number in the offspring from the
mean of the population to which selection was applied) shows a linear
relation to the intensity of selection (i.e. departure of the mean chaeta
number of the parents actually selected from the mean of the population,
expressed in standard measure). The upper line is for selection for increased
and the lower for selection for decreased chaeta number. The differences
in their slopes suggests *prima facie* a difference between the heritabilities
(h^2) in these selection lines, (Based on Clayton, Morris and Robertson's
data quoted by Falconer, 1960).

its simplest; fitness is directly dependent on the phenotype in respect of that character. Under these circumstances we would expect the rate of progress in the character to be proportional to the genetic variance, and indeed the magnitude of the response to selection (as measured by the difference between the means of the offspring and the parental population) relative to the selective differential (that is the deviation from the parental population mean of the mean of the individuals selected and used as parents) is used as a measure of the proportion of the variation that is heritable (see for example, Falconer, 1960). This fraction, that the heritable variation bears to the total, is often called the heritability of the character, though the concept is imprecise as it does not specify whether all heritable variation, including that due to dominance and genic interaction, or just the D-type variation, or any given combination of these components, should be taken into account. Partly for this reason, partly because although all the selected parents have an equal chance of contributing progeny they do not necessarily contribute equal numbers of progeny, and partly because not all of the progeny survive equally well, heritability as measured by the response to selection does not always agree with expectation based on other methods of measuring it. Broadly speaking, however, the response to experimental selection, at least over the early generation of selection, serves to illustrate this aspect of Fisher's principle. An example from a selection experiment quoted by Falconer is shown in Fig. 12.

Disruptive selection

As we have already observed, variation in the environment over the range of territory occupied by a population will lead to variation in the optimal phenotype. This may do no more than introduce a statistical element into the definition of the optimal phenotype, as again we have already observed. Where, however, the differences between two or more of the variant environments, or ecological 'niches', as they are often called, are large enough, sharp enough and permanent enough, the population will encounter selection not towards a single smoothly varying optimum, but towards two or more distinct optima each characteristic of one of the niches. Selection of this kind towards different optima in different groups of individuals must tend to have a disruptive effect on the continuity of variation in the population (Fig. 9). Such disruptive selection can come about in several ways which we will examine together with their genetical consequences in Chapter 8. It will suffice for the moment to note that within each of the groups there will be stabilizing selection towards that group's optimum, and also one must suppose, directional selection when the group means are separating toward their respective optima and when the optima themselves move as a result of those persistent changes of the environment over time whose occurrence we can hardly doubt.

Types of environmental change

In additional to spatial variation of the environment over a population's territory, there must be temporal changes in the environment which the population experiences. These changes over time may be of four different kinds (Fig. 13). In the first place the vagaries of, for example, the weather will ensure that the environment in any one place is not the same at corresponding periods of different years. Yet in the main these changes will represent fluctuations of a random kind from the organism's point of view. To adjust to such fluctuating changes, whether arising for meteorological or any other reasons, would confer no advantage on the individuals since such change is as likely to reverse the direction of its predecessor as to continue it.

The second kind of environment change is cyclical, such as is produced for example by the succession of the seasons during the course of the year. Such changes will of course, have fluctuations superimposed on them, but they are themselves regular rather than random. As such they may be met by direct adaptation of the individual organism. Thus the annual cycle of the seasons may be met by a corresponding annual cycle of development or behaviour; for example the deciduous habit in longer lived plants, the dormant seed stage of annual plants, hibernation in animals and so on. Such

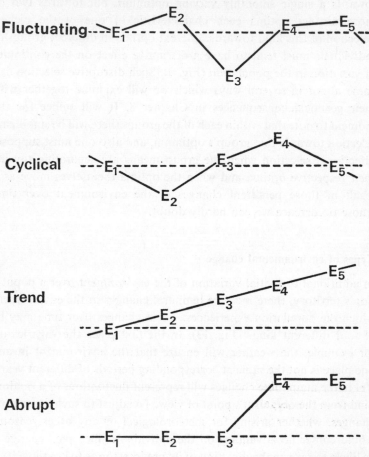

Fig. 13 The four elemental types of environmental change. The environments $E_1 - E_5$ are succeeding one another in time (horizontal axis) and varying from one to another in respect of the optimal expression they determine for the character (vertical axis). In each case the dotted line is drawn through the optimum in E_1 to bring out the nature of the changes.

adjustments will serve and serve well if the full cycle of the environment can be contained in the life cycle of the organism. They will not, however, serve if the life cycle is shorter than the environmental cycle, as it is for example with ephemerals in relation to the annual cycle of change. In such species each generation inevitably must live in an environment different from its parents'. Yet its descendants in the next or some later generation will equally inevitably experience an environment just like that of the original parents, apart from fluctuation. Response in the genetical constitution of the population to the changing selective forces of the environment would thus confer no advantage, for it would always be overtaken by further progression of the environment along its cycle and so would be ineffective. Indeed, in so far as response to selection fixes and so uses up genetical variability, such responses would be actively harmful since they would squander the capital of heritable variability, upon which must depend effective and even essential adjustments to other types of environment change, in fruitlessly chasing an environmental cycle with which they could never catch up. And they can never catch up because selective responses are always a generation late: the genetical adjustment of a population to its circumstances can never come earlier than in the offspring of the parental population which suffered the rigours of selection imposed by those circumstances.

So genetical response to cyclical, and indeed also to fluctuating, changes of the environment are useless in that they achieve no adequate adjustment of the phenotypes and will even be harmful in so far as they use up variability fruitlessly. Longer term changes of the environment, which persist and even strengthen over the generations, are however a different case, for genetical response to the forces of selection which impinge on the parents will lead to better adjustment of the offspring. These longer term changes may themselves be of two kinds, one in which the alteration of the environment continues steadily over a number of generations and so constitutes a trend change, and another where the alteration comes quickly and, if at all large, cataclysmically for the organisms, and so may be described as abrupt or cataclysmic change. This classification of the longer-term changes into trend and abrupt is obviously one of extremes, for one can readily imagine changes of the environment which take the time-equivalent of two or three generations to work themselves out, yet do not continue long enough to be regarded properly as continuing trends. And in as much as the unit of time we are necessarily using is the generation, rather than one independent of biological phenomena like the year, the distinction between trend and abrupt change will vary with the species and the length of its generation time. What is relatively abrupt for a species with a long life cycle could be much more a continuing trend for its shorter-lived fellow.

The shorter term cycles and fluctuations of the environment will be going on side by side not only with each other, but also with the longer term persistent change, and indeed if the latter follows a slow trend it may be difficult to discern behind the fluctuations. Yet they make different demands on the organism: genetical response to non-persistent change is useless or even harmful, but genetical response to persistent change is advantageous or even essential. Once again there is a conflict of demand just as there is between the demands of stabilizing selection and directional selection. Cycles and fluctuations, like stabilizing selection, will encourage genetical invariance (though they may encourage adaptive plasticity of development) while persistent change like directional selection – indeed because they lead to directional selection – will encourage variability. We shall discuss the resolution of these conflicts in the next chapter, but we may note immediately that the balance of impact of the non-persistent and persistent changes which are going on side by side in the environment must depend on the length of the life cycle of any species in question. We have already observed that the annual cycle of seasonal change can be met by a corresponding cycle of development or behaviour of the individual in species whose life cycle is a year in length or longer, whereas this way of coping with them is denied to the ephemeral. In the same way, an annual plant or animal will feel the full impact of fluctuation in the environment from year to year, whereas such fluctuations will have little genetical impact on longer-lived species – perennials as the botanist would call them. We must expect, therefore, that the conflict of demand will impinge differently in species with different life cycles, and that the means by which species resolve this conflict will differ correspondingly.

6 Life-Cycles and Genetic Systems

Modes of reproduction

We saw in the last chapter that the selective forces exerted by the environment on a population can make conflicting demands in two respects. Stabilizing selection favours offspring being like their parents while directional favours them being different. At the same time response of the population to fluctuating changes in the environment and even more strongly, cyclical changes that cannot be accommodated within the organisms life-cycle, is useless and even disadvantageous and hence to be resisted, while response to persistent change is advantageous and even essential. Stabilizing selection and non-persistent environmental change favour genetical invariance in the population, while directional selection and persistent change require genetical variation. How are these conflicts resolved? The answer will inevitably be some form of compromise between meeting the conflicting demands – and a compromise which we must expect to be struck at different points and achieved in different ways according to the life-cycle and biology of the organism. To see how this compromise is achieved we must look at the modes of reproduction available to the different species. These are basically three: asexual, pseudo-sexual, and sexual.

Asexual, or clonal, reproduction is common in plants including higher plants where it takes place by bulb, corm, tuber, tiller, rhizome, stolon, the production of special propagules and indeed in a wide variety of ways. It is much less common in animals, though in lower forms, as in lower plants and bacteria, binary fision and

budding are widespread. It can be a very successful form of reproduction, as is shown for example by the clone of the creeping grass
Festuca rubra which Harberd (1961) found represented over a patch
220 metres long (and see Jones and Wilkins, 1971). Asexual reproduction does not involve meiosis and fertilization, and the progeny
to which it gives rise are therefore invariant in respect of their
nuclear genotype. The phenotype may, however, vary as a result of
changes in the extra-genic parts of the cell. It is well known that
plants propagated continuously by asexual means tend to accumulate
viruses, and the avoidance of such accumulation in domestic plants
like potatoes and raspberries requires special measures. The cytoplasm itself may also change under continuous asexual reproduction,
as has been demonstrated in certain fungi (Mather and Jinks, 1958).
These cytoplasm variants are cleaned out, as are at any rate most
viruses, by a single passage through the sexual cycle. Though a source
of genic variation by segregation and recombination, sexual reproduction is thus a stage of cytoplasmic standardization.

Pseudo-sexual reproduction is a system having the appearance of
sexual reproduction but which omits meiosis or fertilization or both.
In plants pseudo-sexual reproduction may take several forms which
are jointly subsumed under the name of apomixis (as opposed to
amphimixis or true sexual reproduction). Parthenogenesis is one of
the common forms of apomixis in plants and it is the only known
type of pseudo-sexual reproduction in animals (Darlington, 1937).

The precise consequences of this type of reproduction depend on
the form it takes. Sometimes, as in nucellar embryony a normal
diploid maternal cell develops into an embryo without meiosis or
fertilization. The genic consequences are then exactly as with asexual
reproduction: the offspring is like the parent. The same is true if
meiosis is replaced by a mitotic division in megaspore mother cell
or oocyte. Sometimes however meiosis commences but does not
run its normal course: the first anaphase fails and so gives a diploid
restitution nucleus, or two of the four products of meiosis fuse to
restore the diploid number. Some limited segregation may then take
place of a gene heterozygous in the parent, as a result of crossing-over
between the locus of the gene in question and the centromere. This

type of behaviour has consequently been referred to as sub-sexual (Darlington, 1937), but the variation thus appearing in the progeny is very restricted. So, despite such sub-sexual behaviour, pseudo-sexual is basically like asexual, a form of reproduction giving rise to progeny which do not display genic variation. One surmises, however, that at least in parthenogenesis where mother cell or oocyte is formed, there may well be the restandardization of the cytoplasm which takes place when reproduction is sexual but which apparently does not occur when it is asexual, though there is no evidence either to substantiate or to refute this view. If however this is the case, parthenogenesis combines something like the genic invariance of asexual reproduction with the cytoplasmic clean up, including the elimination of viruses, of sexual reproduction.

Sexual reproduction characteristically involves the reduction of the chromosome number by meiosis and its restoration by fertiliza-tion, and these prospectively lead to segregation and recombination of the genes. Segregation and recombination cannot, however, take place in the production of the gametes unless the individuals under-going meiosis are heterozygous at the loci in question. Furthermore heterozygotes cannot be produced, nor can genes be brought together in combinations which did not exist in the parent diploids, unless gametes with dissimilar genes come together in fertilization. We have already seen in Chapter 3 that, if it is effective, inbreeding leads to homozygosis. Sexual reproduction will produce and go on producing variation through segregation and recombination only where at least some heterozygosity is maintained, i.e. where at least some of the fertilizations in the population are between gametes brought together by crossing different parents. We saw this, too, in a different context in Chapter 4.

The flow of polygenic variability from one state to another depends on the presence in the population of heterozygotes produced by crossing and in turn giving rise to segregation and recombination, (see Fig. 7). Free variability and homozygous potential variability are converted into heteɪozygous potential by crossing and are produced from heterozygous potential by segregation, the precise distribution of this outflow between free and homozygous potential being determined by recombination. Furthermore homozygous potential variability can become free only by first passing through the heterozygous potential state. In the absence of heterozygotes, that is in the absence of crossing, the flow of variability dries up, the system becomes frozen and the potential variability cannot be freed to become the raw material for selective action. Crossing is the tap which controls the flow of variability. Thus, by drying up the flow of variability, inbreeding denies to sexual reproduction its special property of producing genically variable progeny, while outcrossing or outbreeding promotes it. Sexual reproduction may therefore produce generally invariable or variable progeny according to the breeding system, or system of mating, which characterizes the species.

Since sexual reproduction with inbreeding leads to genetically uniform progeny because of the homozygosis consequent on the inbreeding, it can operate as a system of invariable reproduction only where the parents are homozygous, and indeed, again as we have seen in Chapter 3, the early generations obtained by inbreeding heterozygotes are marked by segregation, albeit segregation which diminishes in frequency until homozygosis is obtained. There is thus a limit set to the action of this system: in the absence of further special mechanisms it cannot produce from heterozygotes progeny which are uniformly heterozygous like the parent. Nor, of course, since it involves meiosis and fertilization can it faithfully reproduce chromosome complements which, like triploids for example are mechanically incapable of passing through meiosis without disturbance. No such restriction applies to asexual reproduction: only mitotic divisions are involved and any set of chromosomes which is mechanically capable of passing through mitosis will be reproduced faithfully in the asexual progeny, as will any heterozygosity of the genes those chromosomes carry. In the extreme, this means that odd numbered polyploids, notably triploids and pentaploids, can reproduce themselves asexually, whereas sexual reproduction of any kind would be barred to them in the absence of special mechanisms. Indeed, this is true of any chromosome complement which is meiotically unstable. Asexual reproduction has thus allowed populations and species even with triploid and pentaploid constitutions to arise and survive in nature (Darlington, 1937) of which the triploid *Tulipa saxatilis* is an example where the diploid ancestor has been superseded and lost (Darlington, 1963).

Such species as this one, while retaining their flowers as vestiges of the past, must obviously have come to rely on their asexual systems as their sole regular means of reproduction. By far the majority of species which display asexual reproduction, however, combine it with sexual reproduction provided the chromosome complement is not meiotically incapable. Such a combination holds obvious advantages, for no matter how heterozygous it may be, an individual well adapted to its environment can reproduce itself as faithfully as any force of stabilizing selection could require, while it still retains the capacity for producing, via the sexual cycle, the genetical variants on which would depend response to the directional selection imposed by a persistent change of the environment. It is not surprising therefore to find that a great array not only of higher plants, but also of lower ones with both asexual and sexual spores, have adopted this joint style of reproduction. We should note, of course, that in such cases the sexual reproduction must be by outbreeding if it is to form an effective feature of the species.

The same versatility is to be found with pseudo-sexual reproduction. Again, where meiosis has been replaced by mitotic division, or where meiosis though begun breaks down to give a diploid nucleus by restitution, aberrant chromosome complements can be passed on faithfully and uniformly to the progeny. We can see an example of this in the parthenogenetic tetraploid and octoploid brine shrimps of the species *Artemia salina*. As with asexual reproduction, in such a case parthenogenesis has become the only regular means of reproduction. Again however as with asexuality, apomixis is not always obligatory: it can be facultative and join with true sexual reproduction. Such facultative apomixis can combine, in the way that asexuality does, the ability to multiply by pseudo-sexuality a successful genotype, even though this depends on heterozygosity for its success, with the ability to produce by sexual means variant progeny which can provide the basis for response to directional

H

selection. At times it would appear that the two modes of reproduction are adjusted in their occurrence to the pressures of selection.
Thus aphids may be sexual in winter, when conditions are unfavour-
able, numbers small and selective forces presumably strong, but
parthenogenetic in summer when the pressures are relaxed, numbers
rising rapidly and selection presumably weaker. In most cases,
however, the reasons for following one mode of reproduction or the
other at a given time, in a given individual or a given part of the
individual, are not clear. It is nevertheless clear, again as with the
combination of asexual and sexual reproduction, that facultatively
apomictic species are commoner than obligate apomicts, at any rate
in plants (Nygren, 1966): evidently the flexibility afforded by the
combination of invariable and variable reproduction offers advantages, which we can understand when we relate them to the conflicting
pressures of stabilizing and directional selection.

Breeding systems

Sexual reproduction itself can give rise to progeny genetically like
one another and their parents, or to progeny which show genetical
variation. The amount of variation shown will depend on the flow
and distribution of variability, which in their turn depend on the
frequency of outcrossing as opposed to inbreeding, on the occurrence
of segregation and the amount of recombination (Chapter 4).
Crossing, segregation and recombination are thus characters of key
importance to the species and population in their adjustments to the
conflicting demands made on them by stabilizing and directional
selection, and by the need for avoiding undue response to non-
persistent changes of the environment while responding to persistent
changes. It is significant, therefore, to find that they are themselves
subject to genetical control, and hence capable of selective adjustment.

Turning first to crossing, we find, especially in plants, a wide
variety of systems ranging from complete, or virtually complete,
inbreeding to regular, enforced outcrossing. Inbreeding in flowering
plants is commonly ensured by the shedding of pollen from the

anthers before the flower opens. Self-pollination then follows. Thus cleistogamy (in which the flowers never open at all), and the premature dehiscence of the anthers (to be seen in for example wheat, oats, barley and many varieties of tomato where although the flowers open, pollination has generally taken place before they do so), are mechanisms for securing inbreeding. With cleistogamy inbreeding must be complete, but with premature dehiscence out-pollination can occur when the flower eventually does open and it may occasionally be successful, the frequency of success depending on the length of time that has elapsed between shedding of the pollen and the opening of the flower. It is thus not surprising to find that many species, like the cereals, though setting nearly all their seed by selfing, do show a small and variable amount of outcrossing. Inbreeding mechanisms are less well known in animals; but in at least one species, the grass mite *Pediculopsis*, it is secured by a device somewhat reminiscent of cleistogamy, the young becoming sexually mature and mating before they are born, thus ensuring brother-sister mating. Self-mating is also known to occur in the black slug, *Arion ater* (Williamson, 1959) though to what extent it is mixed with crossing appears not to have been recorded.

Inbreeding is secured by pollination within the flower, where obviously it is in any case most likely to take place. Outbreeding can equally be encouraged, though not ensured, by reducing the chance of pollination within the flower, and many devices exist to this end. The stigma may not be receptive at the time when the anthers shed their pollen; this is protogyny where the stigma ripens first and protandry where the anthers do so. The time between the shedding of pollen and ripening of the stigma in *Epilobium* has been found to be capable of change by selection, and these two events can thus be brought to occur together or their order even reversed, protandry becoming protogyny. There must therefore be genetical control and the capacity for genetical adjustment of the rate of outbreeding in this species. In *Nicotiana rustica* it is not just the timing of anthers and stigma that seems to be important, but the heights at which they are positioned inside the corolla tube (Breese, 1959). Their relative positions can be altered by selection and the further apart they are (or the greater the heterostathmy, to use Breese's term) the lower the proportion of seed set by self-pollination and the greater the out-crossing. In the extreme, the anthers and pistils may be separated into distinct male and female flowers, as for example in maize where the male flowers are confined to the tassel at the top of the plant and the female flowers to the ears borne lower down in the same plant. Self-pollination can still occur in such monoecious plants but its chance of occurrence has obviously been reduced, and is commonly reduced still further in maize by protandrous behaviour, the pollen being shed from the tassels before the silks, which bear the stigmatic surfaces, are extruded from the ear.

Such devices as these discourage self-pollination and even prevent it inside a single flower, where it is otherwise most likely to occur; but they cannot prevent pollination of one flower by another of the same plant, which does not of course differ in its genetical consequences from pollination within a single flower. They will thus lead to a combination of pollination by pollen from the same plant (though not the same flower) and from other plants, that is a com-bination of inbreeding and outbreeding, and while, as we have seen, this may itself be favourable, some inbreeding cannot be avoided. Complete outcrossing can, however, be ensured by other devices which are commonplace in both plants and animals. These devices operate by preventing in one way or another the fusion of gametes deriving from the same zygote, which in so far as it promotes random mating among the members of a population is capable in principle of securing the effective maximum degree of heterozygosity (see Chapter 2).

One obvious way of preventing the fusion of gametes deriving from a single zygote is for the male and female gametes always to be borne by different zygotes. Such unisexuality, or sex-separation or dioecy, as it is termed by botanists, is widespread in animals, where in most of the great groups it is a regular feature of all species. It is less widespread in plants, where indeed other devices to which we will turn in a moment are commoner, though it is still by no means rare among them. There is a great wealth of literature on the ways in which the sex of an individual (that is whether it bears male gametes or female) is determined, especially in relation to animals where this device to secure outbreeding is so common. Occasionally it is determined environmentally, as in the marine worm *Bonellia*, where a fertilized egg which falls free develops into a female, but one which falls onto an existing female develops into a male. In the vast majority of cases, however, sex-determination is by genetic means. Most commonly it is by the segregation of a sex-difference which may comprise little if anything more than a single gene (as in asparagus or mosquitoes), a chromosome segment extending at times to include virtually the whole chromosome (as in man where the female carries two large X chromosomes and the male a single X with a much smaller Y chromosome) or even more than one chromosome. (Darlington, 1937; White 1954). In such cases one class of gamete, in man the egg, always carried an X chromosome while of the other class, in man the sperm, half the gametes carry an X and half the Y which has segregated from the X at meiosis in spermatogenesis. Random fusion among eggs and sperm then secures that half the zygotes of the next generation are XX and so of one sex while half are XY and so of the other. The segregation of the X–Y difference at meiosis in one sex is the essential feature of this type of sex-determination. The difference, as we have seen, may range from a single gene to one or more chromosomes; the X and Y may be indistinguishable in size and structure, or the Y may be very different cytologically from the X, even being completely absent as in many orthopterans where the males carry one X but no Y; the heterogametic sex, that is the XY sex, may be the male as in mammals and insect groups like the Orthoptera and the Diptera, or it may be the female as in birds, reptiles and Lepidoptera; the sex mechanism may be very stable or it may be unstable as in the fish *Lebistes* where the effective sex-difference has been transferred by experimental breeding from one pair of chromosomes to another, with the heterogametic sex simultaneously changing from the male to the female; but all these systems work and work satisfactorily. So long as there is the necessary sex-difference, heterozygous in one sex and segregating as a unit difference at meiosis in that sex, the requirements of sex determination are satisfied.

Before we leave this type of sex-determination however one further point must be made. The sex-difference between the X and Y chromosomes switches development into paths leading to males on the one hand and females on the other. That these developmental paths do lead to functioning males and functioning females and not to some sexually aberrant or anamolous type, is also a property of the genotype, though of the genotype as a whole and not just of the X – Y difference. There is a great deal of evidence for this role of the background genotype, but perhaps the most revealing comes from the plant *Lychnis dioeca*. If the two forms of this species, the red and white campions, are crossed, or if either is experimentally inbred, the background genotype is disturbed in ways which we shall discuss in the next chapter, with the result that development is upset and, in particular, the males tend to become hermaphroditic (see Mather, 1948). Thus the genetical mechanism of sex determination comprises two parts: the general genotype which is adjusted to endow the individuals with the capacity for developing into either good males or good females; and the switching mechanism which determines which of the two paths development will follow in any given individual.

Though this is the commonest genetical mechanism of sex-determination, it is by no means the only one. Among the insects, whose versatility in respect of sex-determination as in other genetical and cytological characteristics exceeds that of any other group of higher organisms, the Hymenoptera have a remarkable mechanism in which diploid individuals developing from fertilized eggs, are normally female while haploids, coming from unfertilized eggs, are male. Here sex-determination depends immediately not on segregation at meiosis but on the occurrence or otherwise of fertilization in the other half of the sexual cycle. Thus any of a variety of events in the life-cycle – the environment in which the individual finds itself, chromosome segregation at meiosis, or the occurrence of fertilization – can switch sex-determination: anything, in fact, which can offer a choice between clearly distinct situations would appear to be capable of acting as a switch.

Separation of the sexes will prevent selfing; but it cannot of itself prevent inbreeding in other ways. Indeed we have seen that in *Pediculopsis* inbreeding is secured by sib-mating even though the sexes are separate, and there must always be a risk of such inbreeding wherever individuals tend to be aggregated in family groups. It is not therefore surprising to find separation of the sexes reinforced by other mechanisms such as the production of unisexual broods of offspring in, for example, the fly *Sciara*, or the virtually ubiquitous social prohibition of incest in man. This utilization of man's unique capacity for socially prescribing and enforcing codes of behaviour for control of the system of mating emphasizes in a still further way the basic significance of the mating system for a population and species and the way any available means may be used to regulate it.

Though some plants rely on dioecy as a device for enforcing crossing, incompatibility in one of its various forms (see Darlington and Mather, 1949; Lewis, 1954), is a much more common mechanism among them. Incompatibility depends for its operation on, and derives its name from, the inability of certain classes of pollen grain to germinate on the stigmata or grow effectively down the styles of certain classes of individuals even though both pollen and styles are functionally capable in other combinations. The distinctions among the classes of pollen or of styles are basically genetical in determination though developmental processes may play an important part especially in more complex cases. In the simplest situation (some of whose consequences we have already examined in Chapter 3) incompatibility is determined by a series, and often a very lengthy series (up to 100 or more) of alleles, commonly designated $S_1, S_2 \ldots$ A pollen grain carrying say S_1 will not effectively grow down a style carrying S_1 as one of the two alleles in its nuclei, and so on. This clearly rules out selfing, for although self-pollination may take place it will be incompatible in that the pollen will be ineffective on the styles of the same plant, and other pollinations, especially of close relatives, may also be incompatible. Plants must obviously always be heterozygous at the S locus, and three types of pollination are possible (Fig. 14):— $S_1 S_2 \times S_1 S_2$, which will include selfing and will set no seed as neither S_1 nor S_2 pollen will be compatible; $S_1 S_2 \times S_1 S_3$, which is often described as half-compatible as only the S_3 will be

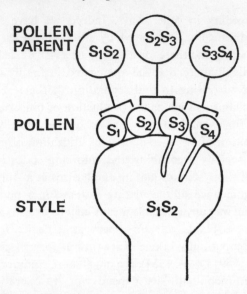

Fig. 14 Incompatibility of pollen and style in plants. The behaviour of the pollen reflects its own genotype in respect of the S gene. It will not germinate and grow successfully down a style carrying the same S allele. All individuals must thus be heterozygotes and each therefore produces two kinds of pollen, both of which may be incompatible (left), half incompatible and half compatible (centre) or both compatible (right) on a given style. All self-pollination is of the first type and incompatible.

compatible, but which will nevertheless give a full set of seed if the pollen supply is adequate; and $S_1 S_2 \times S_3 S_4$ in which all pollen is compatible. In such incompatibility, which is found for example in cherries, *Oenothera organensis*, and in members of the Scrophulariaceae and the Solanaceae, the behaviour of a pollen grain depends on its own genotype in respect of S and the incompatibility is said to be gametophytic in determination.

In other cases, to be found for example in the Cruciferae and Compositae, the behaviour of the pollen is determined by the genotype of the plant that bore it, acting no doubt through the agency of the cytoplasm with which it endowed the pollen grain. Pollen will not grow effectively down a style of the same genotype as that from which the pollen derived. All the pollen grains from a plant will thus act alike in respect of incompatibility and half-compatible matings are impossible. In particular, however, $S_1S_2 \times S_1S_2$ will be incompatible and selfing is thus impossible, as of course will also be certain crosses especially among close relatives. Pollen behaviour in such incompatibility is said to be sporophytic in determination.

Sporophytically determined incompatibility may depend solely on physiological properties, as does gametophytic incompatibility, without any associated differences of form or structure in flower, stigma or pollen. It may on the other hand be associated with morphological differences, in the form of heterostyly as we see in pin and thrum flowered plants of the primrose. In the primrose, as in many other plant species, the stigma and the anthers are borne at different levels within the flower, some plants having their anthers higher than their stigmata (thrums) but with others having anthers at the lower level and stigmata at the upper (pins). There are also characteristic differences in pollen size and stigmatic papillae associated with the pin-thrum difference. These morphological differences are clear and striking, but their significance is far from clear. They have been interpreted as structural adaptations to promote crossing, but even if this is the case its effect can only be secondary, for the decisive difference between pins and thrums lies in a physiological incompatibility reaction (Fig. 15). Pin pollen is functional only on thrum styles and thrum pollen on pin styles, while pin pollen is ineffective on pin styles whether of the same or a mother pin plant, and thrum pollen is ineffective on thrum styles, as Darwin established experimentally a hundred years ago. The genetical determination of the difference between pin and thrum plants is simple, depending on a single gene difference, pin being ss and thrum Ss. The incompatibility reaction ensures that all successful pollinations are between ss and Ss, and the system is thus self-perpetuating like the X – Y system of sex-determination.

A more complex form of heterostyly is known in *Lythrum salicaria* and *Oxalis valdiviana*, where there are three levels in the flower, two being occupied by anthers and the third by the stigmata. There are thus three types of individual, differing in the level at which they

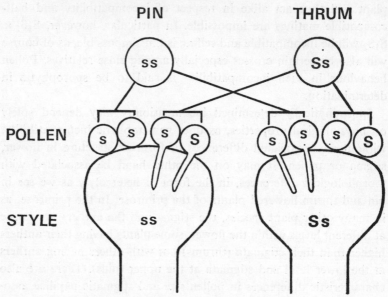

Fig. 15 Incompatibility in a heterostyled plant such as the primrose. The behaviour of the pollen reflects the diploid genotype of the parent that bore it. Pin pollen, all of which is s, will not grow down a pin style but is compatible on a thrum style. Thrum pollen is half S and half s, yet nevertheless behaves all alike and will not grow down a thrum style but is compatible on a pin style.

carry their stigmata and compatible pollinations are possible only between anthers and stigmata at the same level. The three types of individual are determined by genic differences at two loci.

In all these types of incompatibility the switching of the system depends on simply inherited differences among alleles at a single S locus or occasionally two loci. It is known, however, that although simply inherited, the differences among the alleles at the S locus are compound (see Lewis, 1954). This is especially clear in the case of heterostyly where, although normally transmitted as a unit in inheritance, the S–s differences comprises a number of parts each associated with one or other of the morphological or physiological differences between pins and thrums. Evidently the switching gene has been built up in evolution, presumably by the action of selection, to give the smoothly functioning mechanism that we see today, and observations on the Chinese primrose (*Primula sinensis*) show us how this may have come about (Mather, 1950). Furthermore, as in the case of sex separation, the background genotype is adjusted to give the incompatibility mechanism in all its precision of operation, which the S genes switch. It can be broken down by inbreeding, by over-wide crossing and by selection, so that the switching genes though still present are associated with incomplete, imprecise and inefficient incompatibility reactions, or even with none at all (Darlington and Mather, 1949). And morphological differences as well as the incompatibility reaction may be altered by change in the background genotype, as has been observed by Ornduff (1972) in his study of the breakdown of tristyly in *Oxalis*. Since the basic mechanism can be broken down selectively, presumably by disruption of the polygenic background, it is to be concluded that it was built up by adjustment of this genetical background under the action of selection in the course of evolution.

Incompatibility is not confined to the flowering plants. It is widespread in the fungi, where it is known as heterothally and operates at various levels of complexity, depending on two or more alleles at one or two loci. In the basidiomycete *Schizophyllum commune* these loci are known to be structurally complex, and it is also known that each of the two has a very large number of alleles with a world-wide distribution (Raper *et al.*, 1958). Nothing is, however, known of the importance of the genetic background for the operation of heterothally in fungi. A simple case of heterothally has been demonstrated in the ferm *Pteridium aquilina*, (Wilkie, 1956), but nothing is known of it beyond its dependence on two alleles at a single locus.

Thus incompatibility reveals the same broad features as sex-determination. It displays a variety of switching mechanisms: gametic determination and sporophytic; with accompanying morphological differences and without them; dependence for its operation generally on protein differences but in at least one species, *Linum grandiflorum*, on the osmotic relations of pollen and style; with switching depending sometimes on the segregation of two alleles and sometimes on that of many. The switching genes are simply inherited but are structurally complex, sometimes as in heterostyly very complex, and as their capability for break-down reveals, they must originally have been built up during the evolution of the system. Incompatibility also shares with sex-determination its dependence on the polygenic adjustment of the background or common genotype for endowing individuals with the developmental capacity for being switched into one precise pattern of behaviour or another. Again, this adjustment of the background genotype can be broken down by undue inbreeding or over-wide outcrossing. It too must therefore have been built up in the past, just as it is maintained in the present, by the action of natural selection. All these types of incompatibility like all the types of sex-determination agree in one respect. They prevent self-mating, which without them would be the most likely kind of inbreeding to occur as well as the kind leading most rapidly to homozygosis. They can be understood only as adaptive mechanisms, built up in their various ways by selection for control of the breeding system.

Outbreeding, by stimulating the flow of variability and hence genetic flexibility, enables populations to respond to directional selection. In so far, therefore, as the breeding system encouraging outbreeding is itself a genetically controlled and adaptively adjusted character, it, like other characters, can respond to any change in circumstances and in particular to any change which reverses the advantage of outbreeding and favours inbreeding. It is not uncommon, in fact, to see inbreeding mechanisms superimposed on outbreeding devices. Premature dehiscence of the anthers in cereals secures inbreeding despite the obvious adaptation of the flower to wind pollination, which would otherwise lead to crossing. The sinking of the stigma inside the cone of stamens in the tomato has a similar effect. We have already seen too how inbreeding is secured by enforced sib-mating, despite separation of the sexes, in *Pediculopsis*. Turning to heterostyled species, in some populations the primrose produces individuals with both anthers and stigmata at the same high level in the flower, and self-pollination in such long homostyles is as legitimate as crossing is between pins and thrums (see Ford, 1971). In species with incompatibility depending on the action of a series of alleles at the S locus, many cases are known of individual alleles which are ineffective in mediating incompatibility or even capable of over-riding it in their fellow alleles, so permitting and even encouraging effective self-pollination. The self-fertile garden species, *Antirrhinum majus* is claimed still to possess the S locus, with its alleles producing gametophytically determined incompatibility, which one finds in its close relatives like *A. glutinosum*; but unlike its relatives *A. majus* is capable of effective self-pollination because the action of its S alleles is suppressed by a non-allelic gene, for which it is homozygous and which its relatives appear not to possess.

In all these cases we can see the history of change because of the stratification of the breeding system, with the new inbreeding overlaying the formerly effective outbreeding device. The change has been effected not by undoing the outcrossing mechanism but by vitiating its action sometimes by alterations of the S genes, at others by the action of a non-allelic gene. Again we see the same end being achieved by more than one genetic means: so long as the move towards inbreeding is secured the genetic means of securing it is of secondary importance.

Outbreeding, by securing genetical flexibility, provides the genetical basis for its own suppression through the rise of inbreeding devices, if circumstances change. Inbreeders are in a different case for by freezing up the flow of genetic variability and rigidifying the genotype, inbreeding denies to its practitioners the means of developing outbreeding devices through selective adjustment. The stratified breeding systems that we see are therefore either inbreeding superimposed on outbreeding, such as we have just been discussing, or outbreeding devices reinforced by further outbreeding devices like the unisexual broods of *Sciara* and the suppression of incest by social action in man that we referred to on p. 111. And if inbreeding cannot be superseded or replaced by outbreeding, the inbreeding must be an evolutionary dead-end because of the genetical rigidity to which it leads, denying the ability to respond to directional selection. Not all inbreeding devices should, however, be viewed as stages on the road to final rigidity and extinction: they may be no more than checks on excessive outbreeding in the population. Thus, despite the common occurrence of homostyles (p. 155) in the primrose populations of certain parts of Southern England, no population is known to be wholly homostyle. In other words, no population is wholly in-breeding, wholly homozygous and wholly rigid, though the populations where homostyles are common must be presumed to be more inbred, more homozygous and more rigid than the normal hetero-styled populations, just as self-compatible species of *Leavenworthia* show less variation within populations than do species where the possibilities of inbreeding are more restricted (Solbrig, 1972). There is in such cases, as in other cases mentioned earlier, a prospective compromise to be achieved between inbreeding and outbreeding and the balance should change as the relative impacts of the two types of selection, stabilizing and directional, change. And so long as any crossing occurs at all, and at least some genetical flexibility is main-tained thereby, reversion to complete outbreeding is as possible as further progression towards complete inbreeding.

Segregation and recombination

Crossing is an essential step in ensuring a flow of variability and the genetical flexibility on which response to directional selection depends. Crossing between genetically unlike parents leads to heterozygosity and with random mating heterozygosity for any pair of alleles is maintained at its effective maximum of $2uv$ in the population where u and v are of course the frequencies of the alleles. The variability is, however, redistributed by the segregation of the alleles from the heterozygote and in a polygenic system it is further controlled by the frequency of recombination among the various member genes of the system (Fig. 7). We must, therefore, now turn to look at the properties and control of segregation and re-combination.

Normally segregation is an inevitable consequence of hetero-zygosity and leads to the haploid products of meiosis carrying the two alleles with equal frequencies. The alleles may, of course, fail to be recovered equally often in the progeny to which these haploid products lead, because of viability or competitive differences among the resulting zygotes, or even among the haploids themselves, at any rate in plants; but these are no more than modifications superimposed on an originally normal segregation. Sometimes however, albeit rarely, the original segregation may be distorted. In fungi, the production by heterozygotes of tetrads showing aberrant ratios with one or other allele present in three or four of the spores and the other in only 1 or even 0, is well known. Such gene conversion, as it is termed, is nevertheless an infrequent event and would hardly disturb the 1 : 1 ratios of the allele in the spores produced overall. Non-disjunction of a pair of chromosomes at meiosis could distort segregation too; but again this is a relatively rare event in normal individuals.

Much more drastic distortion of segregation are known in other cases. In males of the fly *Sciara* meiosis is so modified that the entire paternal chromosome set is eliminated and only chromosomes, and therefore genes, that the male obtained from its own mother are handed on to its progeny (see White, 1954). Males of *Drosophila pseudoobscura* that carry a particular sex-linked gene termed 'sex-ratio' produce a great, though variable, excess of males over females in their progeny, as a result of aberrant behaviour of their X and Y chromosomes. This is perhaps a particular example of the phenomenon called meiotic drive, which is known in some strains of *Drosophila melanogaster* and elsewhere and which shows itself by distorted segregation ratios usually with the aberrant allele in excess.

Genes showing meiotic drive occur in wild populations. Segregation Distorter, which is of this kind, in *Drosophila* was discovered in flies from the wild, and the so called *t* alleles are a regular feature of wild populations of mice. Such a gene would soon become fixed in a population if its advantage in segregation was not offset by a countervailing disadvantage. The *t* alleles have several effects. In combination with another mutant allele (T) at the locus they affect the length of the mouse's tail, and can be detected by this test. They show meiotic drive in heterozygous males, though they segregate normally from the normal allele in heterozygous females. But when homozygous they are either lethal to the zygote bearing them or make it sterile. It is presumably the balance between meiotic drive and lethality that preserves them floating in the populations. The mechanisms which result in meiotic drive are unknown. The mechanism which results in the effective suppression of segregation in species of the plant genus *Oenothera*, on the other hand, are well established. (Darlington and Mather, 1949). As we have already noted in Chapter 2, chromosome interchanges are common in this genus, and in many species all individuals are heterozygous for interchanges which result in all, or most, of the chromosomes forming a ring at meiosis. Balanced gametes are found only when alternate chromosomes of the ring pass to the same pole at first anaphase of the reduction division, with the consequence that recombination of their chromosomes is effectively suppressed. The plants therefore breed as if heterozygous for a pair of compound allelic genes, or super-genes, which are termed complexes. Though pollen is formed carrying both of these complexes, only one of them is successfully transmitted to progeny through the male gametes. In the female side, however, the eggs may be found to be carrying only one of the two complexes as the single functional egg of the female gametophyte is formed from one of the two spores carrying this complex, out of the tetrad which is the immediate product of meiosis (the so-called Renner effect). Since the complex which is regularly transmitted through the egg is the opposite of that successful in the pollen, the progeny will regularly be once again heterozygous for the two complexes and, of course, for the interchanges which by their effects at meiosis hold the complexes together. It only remains to add that these species of *Oenothera* regularly self-pollinate, for us to see that they provide us with the paradoxical phenomenon of a true-breeding hybrid, which is regularly heterozygous for many of its genes, but in which segregation is effectively suppressed and which has therefore combined the genetical rigidity of the inbreeder with the hybridity normally associated with outbreeding.

Despite these examples, however, distortion of segregation and in particular controlled distortion, must be regarded as a rarity. It is either exceptional among the genes of the species, as in *Drosophila* and mice, or the species which show it regularly, as in *Sciara* and *Oenothera*, are clearly exceptional, special cases. Only in these special cases it would seem, has the control of segregation proved to be a reliable means of managing variability. When on the other hand we turn to recombination we see a different picture (see Bodmer and Parsons, 1962, for an extensive review).

Recombination can come about in two ways: genes borne on different chromosomes normally assort independently of one another at meiosis and so recombine freely; and genes which are borne on the same chromosome recombine if crossing-over takes place within the chromosome between their respective loci. Genes on the same chromosome between whose loci a chiasma regularly forms show 50% recombination and so recombine as freely as if they were on separate chromosomes. Each chiasma is therefore equivalent in this case to dividing the chromosome into two units of recombination, and Darlington (1939) has consequently defined a 'recombination index' as the haploid number of chromosomes plus the average number of chiasmata in the nucleus, to provide a measure of the freedom of recombination.

The frequency of recombination among the genes of a genotype can thus be altered by change in the number of chromosomes or in the frequency of chiasma formation, providing the two types of change do not occur simultaneously in a self-cancelling fashion. Obviously the number of chromosomes is, apart from the special case of B chromosomes, a stable heritable character; but it can nevertheless change, not only by polyploidy, but by reassociation of the chromosome material within a haploid complement. Such reassociation involves effective gain or loss of centromeres, which must restrict its freedom of occurrence. It has nevertheless taken place not uncommonly during evolution as we can see, for example, by comparing the number of chromosomes of *Drosophila* spp. (see Dobzhansky, 1951).

The frequency of chiasma formation, like other aspects of chromosome behaviour, is under genetic control (see Rees, 1955; Lewis and John, 1963). In rye, for example, inbred lines regularly have a lower and more variable number of chiasmata in their pollen mother cells than do the F_1s between them. Inbreeding from the F_1s results in a decline in the average number of chiasmata, which thus display

I

inbreeding depression (see Chapter 7) in a novel way; but the different inbred lines that separate from one another at F_3 and later generations reflect their differences in genetical constitution by showing differences in the chiasma frequencies they display. The indications are that chiasma frequency is mediated by a polygenic system. To take but one further cytological example, lines of *Lychnis dioeca* taken from the wild, and their F_1s, commonly show characteristic differences in chiasma frequency when grown together in the experimental field (C. W. Lawrence, quoted by Mather, 1960).

The genetical, as distinct from cytological, evidence for the genic control of crossing-over has been reviewed by Simchen and Stamberg (1969). There is extensive evidence from groups ranging as widely as bacteria, fungi, insects and higher plants, testifying to the importance of the genetic background in determining the frequencies of recombination, and in some cases specific genes have been indentified as having such an effect. There is evidence, too, that recombination values can be altered by selection in *Drosophila melanogaster*. This was shown many years ago by Detlefsen and Roberts (1921) for sex-linked genes and more recently Kidwell (1972) has increased the recombination between the genes Glued and Stubble in chromosome III of *Drosophila* from 15% to 30% by selection. The pattern of response over the generations was very similar to that observed with the more familiar external phenotypic characters of the fly (Chapter 4): the increase was achieved in 12 generations after which response effectively ceased; the response in recombination value was accompanied by the correlated response of reduced fertility, so familiar from selection of external phenotypic characters (Chapter 7); and there was evidence of dominance of the genes mediating recombination frequency. In other words this cytological character behaved just as do the familiar external characters like size and chaetae number which are known to be under polygenic control. Selection for reduced recombination was not always effective, but reductions to below 10% were obtained in a number of lines.

Not only is the frequency of chiasma formation, with its concommitant crossing-over, of importance in determining the pattern of crossing-over; but so also are the positions in which the chiasmata tend to form within the chromosomes. In some species, like the liliaceous plant *Fritillaria meleagris* and the grasshopper *Stethophyma grossum*, the chiasmata are almost entirely localized in their formation to positions near the centromeres of the chromosomes while in rye and newts they tend to form near the ends of the chromosomes away from the centromeres. In many other organisms, including lilies, tulips and most grasshoppers, they appear to be more evenly spread along the chromosome arms, though it would seem likely that even in such cases their distribution is not wholly at random (Mather, 1940). Now where a chiasma is closely localized in its position it will effectively divide the chromosome into two segments between which recombination is free but within each of which it is rare (Fig. 16).

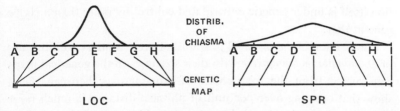

Fig. 16 The effect of localization of chiasmata on the linkage relation of the genes. A – I denote gene loci which are equally spaced along the cytological chromosome (above). When the chiasma is more localized in its distribution along the chromosome (LOC), there is high recombination between the central loci D – F but none among the end loci A – C and G – I. The result is a genetic map (below) in which most of the genes tend to aggregate into two groups within each of which there is little or no recombination but between which recombination is free. Where the chiasma is more spreading in its position along the chromosome (SPR), crossing-over may still not be equally frequent in all segments of the chromosomes but all the genes show some recombination with one another.

Genes within a segment will thus tend to stick together, the whole segment being transmitted much as a unit in heredity, and there will be little release of variability where this depends on reassociation of such genes. A more even distribution of the chiasma along the chromosome will be accompanied by more variation from one mother cell to another in its position, and hence by a more even distribution of recombination among the genes. Fewer of the genes will tend to be locked together regularly in transmission from parent to offspring and the redistribution of variability will then be greater. So not only adjustment of the number of chiasmata but adjustment of their properties in position determination can also affect the flow of variability.

That localization is under genetic control is shown by the cross between the two species *Allium fistulosum*, which has its chiasmata localized near the centomeres, and *A. cepa*, which has them localized near the ends of the chromosome arms (Darlington, 1939). The F_1 resembles *A. cepa* in its behaviour, but in F_2 types like *A. fistulosum* reappear. Furthermore on F_2 plants having bivalents with more evenly distributed chiasmata are also to be found, showing that not only the sites of localization but also the occurrence of strict localization itself is under genetic control and control by more than a single gene at that.

Comparison of the physical positions of genes on the chromosomes of *Drosophila melanogaster* with their spacings on the genetical maps constructed from the frequencies of crossing-over between them, show that crossing-over per unit of physical distance is much rarer near the centromere than it is nearer the middle of the chromosome arms (see Mather, 1938). Evidently there is some localization away from the centromeres in this fly. The fact that the frequency of recombination between genes near the centromere is more variable and more prone to be affected by environmental factors such as change in temperature, than it is between genes further away, suggests that the variation is more likely to be the result of shifts in the position of the chiasmata than in their frequency. It is therefore significant that the genes Glued and Stubble, used by Kidwell in selecting for changed frequencies of recombination, straddle the centromere and that her analysis showed the changes brought about by selection to be greater near the centromere than further away from it. Thus it may well be that the chief effect of selection in this case was on the position of chiasma formation rather than on its overall frequency.

Since the properties of chiasma formation and crossing-over are under the control of genic, and one suspects commonly of polygenic, systems they are capable of adjustment by the action of natural selection on these systems. They may be altered in other ways too. Crossing-over may be abolished altogether as it is in the male *Drosophila*, recombination within chromosomes being confined to the females and so reduced to half the value it would show if both sexes showed a normal pattern of meiosis. The abolition of crossing-over on both sexes would of course simultaneously abolish all recombination within chromosomes. This is a general effect on all parts of all the chromosomes, but more restricted effects on recombination are produced by inversion heterozygosity. Where inversions are floating in a population, as is not uncommon (see Chapter 2), the exchange of genes will be rare between the relatively inverted sequences, especially near the ends of the inversion, and there will thus be blocks of genes capable of recombining in individuals homozygous for one sequence or the other but not of doing so between sequences. Adjustment of these genes can take place only within sequences which may thus become partly isolated from one another in the development of the genic combinations they carry. Some of the selective consequences of this will be taken up in the next chapter.

Inversions restrict recombination within chromosomes. Interchange heterozygosity goes further and restricts it between chromosomes. Taken to its limit, this results in the locking together of the genes of a whole haploid set of chromosomes into a single unit of transmission from the heterozygote as we have already seen in certain species of *Oenothera*. It is this ability to restrict recombination and bind blocks of genes together into almost permanent units or super-genes as Darlington and Mather (1949) have called them, that gives inversions and interchanges their significance in populations.

Genetic systems

In considering how populations and species can meet the conflicting demands made on them by stabilizing and directional selection, and by persistent and non-persistent selective forces, we have recognized three basic types of reproduction, asexual, pseudo-sexual (or apomictic) and sexual. Asexual and pseudo-sexual are essentially modes of reproduction which omit meiosis and fertilization and they can therefore perpetuate in reproduction any chromosome and genic complements which can pass successfully through mitosis whether they are meiotically capable or not. But even with genetic complements capable of passing successfully through meiosis, asexuality and pseudo-sexuality also offer the opportunity of reproduction without variation, because they avoid reduction and its concommitant segregation. Sexual reproduction is, however, more versatile, for it can offer reproduction without variation or with it according to the pattern the sexual cycle follows. In further exploring this versatility, we have recognized two further systems, the breeding system on which generally depends the degree of heterozygosity of individuals in the population, and the chromosome system on which depends segregation and recombination in the progeny of heterozygotes. These three systems, reproductive, breeding and chromosome, may be regarded as together making up the genetic system of the species as Darlington (1939) has called it. On the genetic system depends the type and degree of variation of the progeny (i.e. the extent to which the offspring resemble their parents and one another) or to put it in the terminology of Chapter 4, the flow of variability which in its turn will govern the response of the population to such selective forces as may impinge on it.

Only asexual and pseudo-sexual reproduction by their avoidance of normal meiosis can offer the opportunity of perpetuation to aberrant types like unusual polyploids and wide hybrids. Equally, on the other hand, only sexual reproduction can offer the opportunity of producing widely variable progeny, though the extent to which sexual progeny show variation is subject to further control by adjustment of the breeding system and the chromosome system. Since these two systems are themselves subject to genetic control, the degree of variability may be adjusted by the action of selection to give invariability, wide variability or any compromise between these two extremes. Compromise between invariability and wide variability may also be reached by combining asexual and sexual, or pseudo-sexual and sexual reproduction, provided of course that the chromosome and genic complements are meiotically, as well as mitotically, capable. Thus several kinds, as well as many levels, of compromise are possible, each with its own special features and properties, and the one achieved might be expected to reflect both the selective circumstances of the population and the biological features of the species. Long-lived species might be expected to show little but sexual reproduction with outbreeding and reasonably free recombination. Ephemerals on the other hand, under the pressure of cyclic change in their circumstances might be expected to compromise at a low level of variability, whether this be achieved by a high-rate of inbreeding as in many plants, or by outcrossing with severe restriction of recombination as in *Drosophila*. A persistent environment like that of a woodland or grass-sward, might be expected to favour asexual reproduction for propagation and local spread, combined with outbreeding sexuality for colonization, while the weed of arable land with its continual change of environment should find preponderant inbreeding, with variable dormancy of the seed, of more advantage than asexual propagation for combination with some outbreeding.

Whatever the circumstances, however, the capacity of the genetic system for meeting them must be judged as a whole, for though formally separable in discussion of their control and consequences, the three sub-systems, reproductive, breeding and chromosome, overlap and interact in their working and are in some measure capable of reinforcing or substituting for one another. The facultative apomict could find little advantage in its sexual reproduction if this were by inbreeding, and in circumstances imposing heavy stabilizing selection, the preponderant inbreeder could afford a less severe restriction of recombination than the preponderant outbreeder. And *Oeneothera spp.* could not be the true-breeding hybrids that they so often are, if their system did not combine inbreeding with severely restricted recombination and the control of effective segregation. The genetic system works as a whole and must be adjusted as a whole, with the corollary that its sub-systems though being subject to individual adjustment must be adjusted in step with one another if the counterpart of the fate that overtook the experimental *Antir-*

rhinum plants of Mather and Vines (1951), which combined self-incompatibility of pollen and style with cleistogamy of the flower, is to be avoided.

The genetic system that we see in a species is the system that has evolved under the selective pressures of the past, whether natural in the wild or artificial in our domesticated plants and animals. It has served the species satisfactorily in the past and will do so in the future if the selective pressures to come are like those under which it evolved. If however they are not, either the genetic system must readjust or increasing unfitness and prospective extinction of population, and even whole species, will follow. Being a character like any other, even though it is at the same time both of supreme importance to all the other characters and subject to the action of selection only through its effects on the variability of these other characters, the genetic system can and will readjust if it displays the heritable variation on which response to selection depends. Some genetic systems, the completely asexual dependent, the obligate apomict, and the complete inbreeder, will not show such variation and will consequently be incapable of readjustment: the systems may have been essential, each in its own way, to the survival of the odd polyploid or the wide sexually-sterile hybrid, or the inbreeder forced into high adaptation to its specialized environment, but their success has only postponed, not avoided, the ultimate day of reckoning. The outbreeder, on the other hand, is as genetically flexible and prospectively versatile in respect of its genetic system as of any other character. Nor need it be an unrestricted outbreeder for, as we have seen, compromise can be struck under the pressures for present stability and future change: so long as any outbreeding, segregation and recombination persist, readjustment remains a possibility. Genetic invariability may be the key to the maximum success of the present, but outbreeding is the continuous thread running through the varied pattern of the genetic systems that we see around us.

7 Genic Balance and Genetic Architecture

Heterosis and balance

When individuals of a normally outbreeding species, like maize or sugar beet or *Drosophila* or poultry, are made to inbreed by suitable matings, the lines so produced move towards homozygosis at rates which depend, *inter alia*, on the mating system used (see Chapter 3). At the same time the individuals in these inbred lines characteristically show a decline in their vigour and fertility until, as complete homozygosity is approached and attained, they stabilize at their reduced levels of vigour and fertility (Fig. 17). This phenomenon is known as

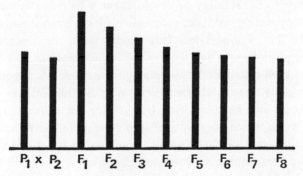

Fig. 17 Heterosis and inbreeding depression in maize. The relative heights of two inbred lines (P_1 and P_2), the cross between them (F_1) and seven generations ($F_2 - F_8$) derived by self-pollination from this cross. Heterosis is displayed by the excess in height of the F_1 over both parents, and inbreeding depression by the progressive reduction in height under the continued selfing. (Results from East and Jones quoted by Darlington and Mather, 1949).

inbreeding depression. The amount of depression varies from one line to another, in some cases being so great that the line itself is extinguished. Within a line the fall from one generation to the next parallels the fall in heterozygosity or, to put it the other way, the rise in homozygosis. Similarly when two such inbred lines are intercrossed, the F_1 characteristically shows a great increase in vigour and fertility. This is hybrid vigour or heterosis. Broadly speaking the F_1 rises to the level of the population from which the inbred lines were begun, though the precise level will vary from one F_1 to another, with some even exceeding that of the original population. If F_2s are raised from an F_1 and the inbreeding resumed, inbreeding depression will again set in. Inbreeding depression and hybrid vigour may be shown by a wide range of characters of the organism, including chromosome behaviour as indeed we have already had occasion to note. Inbreeding depression is also commonly reflected in increased variability not only among the members of inbred lines but also among their repetitive parts like bilateral characters in animals and floral morphology in plants, while heterosis is similarly reflected in an increased uniformity of development among the members of an F_1 and their repetitive parts, so indicating a decrease in the stability of development with rising homozygosis.

Thus inbreeding produces progressively less vigorous and developmentally less well-adjusted individuals: with the restoration of heterozygosity by crossing, vigour and adjustment are also restored. This has been interpreted as springing from an innate superiority of heterozygotes as such (Lerner, 1954) with the implication that the more genes are heterozygous the greater the superiority. Even leaving aside the consequent question of why an individual heterozygous for a pair of alleles should regularly be superior to both of the two corresponding homozygotes in vigour, fertility and developmental adjustment, the wider evidence does not support this view. Even in an outbreeding species like *Drosophila*, different homozygotes are not all alike, and indeed range widely from very poor types to others which are as vigorous as some heterozygotes. Gene content must therefore be important as well as heterozygosity. Furthermore vigour, as expressed in viability, has been shown not to be proportioned to the number of chromosome segments present in the heterozygous state (Breese and Mather, 1960); and not all characters show inbreeding depression and heterosis – the number of sterno-pleural and abdominal chaetae, for example, in *Drosophila* do not. A different explanation must be sought.

Now, naturally inbreeding species stand in clear contrast to out-breeders. Though homozygous, the individuals of such species are vigorous, fertile and developmentally adjusted. When such homo-zygotes of different genotypes are crossed, they may show some heterosis; but it is never as dramatic as in outbreeding species, it may not occur at all and, indeed, the heterozygote may be inferior to its parents. This is well illustrated by Williams and Gilbert's (1960) observations on yield of fruit in the tomato, which in this country is an inbreeder. They found that although F_1s regularly exceeded their better parents where these were low-yielding varieties, the situation was different with higher-yielding parents. Of the 10 F_1s they raised from crosses among five of these higher-yielding varieties, four exceeded their better parents in yield, four fell short and two gave the same yield. Clearly heterosis is not the regular phenomenon in the inbreeding tomato that it is in the outbreeding maize, where F_1s always show considerable heterosis in respect of yield. The tomato provides a particularly clear example, but other cases, perhaps less extreme, could be cited to illustrate further this low level and capricious occurrence of heterosis in inbreeding species. Clearly heterozygosity does not of itself produce heterosis.

Outbreeding and inbreeding species differ in that in the one the individuals comprising a population are heterozygous for many of their genes, while in the other they are homozygotes in the main. The two kinds of species agree in that in both the naturally occurring genotypes give rise to vigorous, fit individuals, as indeed they must if they are to pass the test of natural selection. Thus vigour, adjust-ment and fitness are the properties of genotypes which have passed the test of natural selection and which have survived because they possess these properties – because, we might say, they are balanced genotypes. Why then should outbreeders give poor unbalanced genotypes when inbred, but inbreeders give on the whole good, adequately balanced, genotypes when crossed? We should note first that the normal balance must be achieved in different ways in inbreeders and outbreeders. In inbreeders the balance must be internal to the haploid set of chromosomes which, because of the inbreeding, is matched with its exact counterpart when the diploid zygote is formed. In the outbreeder, this internal balance of the haploid set is of secondary importance, because as a result of outbreeding it is joined at fertilization with a genically dissimilar set, with the consequence that the overall or relational balance of the two sets is the important thing.

Now where a polygenic system is involved, a similar internal balance could be struck in various ways in different haploid sets, and subject only to any dominance and epistasis not being grossly biased in any one direction, the overall balance in the F_1s would be much the same as in the homozygous parents. Only following recombination in the production of the F_2 would the balance be seriously upset (Fig. 18). With the relational balance of outbreeders, however, the situation is different. A successfully balanced genotype heterozygous for no more than 10 gene pairs affecting the character, could give rise on inbreeding to over 1000 genetically different homozygotes, most of which would be unbalanced in greater or lesser degree and many of them grossly so. Furthermore since virtually any pair taken at random from among the inbred lines would be unbalanced in different genetic ways, crossing them would tend to re-establish the balance and so produce heterosis. Relative fitness would tend broadly to follow heterozygosity but would be related also to genic content, and the range among the different homozygotes produced by inbreeding would, as has been observed, be greater than among the heterozygotes from which they were descended and which could be reconstituted by crossing. Moreover, given that inbreeding species have evolved from outbreeders (see Chapter 6), the move towards inbreeding would involve, and indeed owe its success to, a gradual elimination during inbreeding of the many lines of descent leading to poor internal balances, in favour of the relative few that could produce a balance comparable with the relational balance of the outbreeder. Meanwhile, as the move

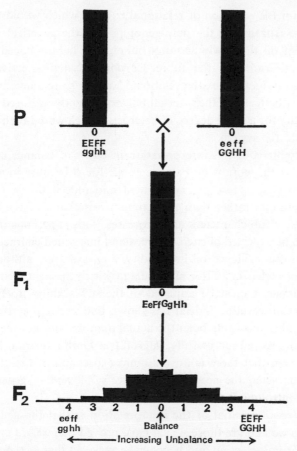

Fig. 18 Loss of balance in F_2 following crossing in an inbreeding species. Four genes (E – H) of equal effect and without dominance or linkage are assumed to be involved, the best balance being achieved with equal numbers of increasing alleles (capital letters) and decreasing alleles (small letters) in the genotype. The two true-breeding parents and their F_1 are all fully balanced, but balance is broken down by segregation and recombination in F_2. The unbalance increases as the number of small letters (left) or capital letters (right) in the genotype rises over equality.

towards inbreeding would presumably be gradual, there would be a continuing, if diminishing selection in favour of relational balance in the partial heterozygotes, and this would mean that the ultimate homozygotes would have been selected not only for internal balance but also for the retention of relational balance which would thus be as it were frozen into the new genotypes by the genetical rigidity consequent on the newly acquired inbreeding habit. Indeed, if the greater uniformity of the inbreeder compensated selectively for some loss of vigour, etc., the relational balance so retained might be sufficiently better than the internal balance recently acquired to give the capacity for a small degree of heterosis, which would then remain frozen into the system.

The essentials of this interpretation in terms of balance are two. The first is the notion of balance itself, for this must imply that maximum vigour, fertility, adjustment and indeed fitness depends on an optimum rather than a maximum expression of the various contributary sub-characters and processes. This is perhaps no more than is to be expected of such complex and integrated characters and there is some evidence of its validity. For example, although the number of piglets in a litter must be a major component of fitness in these animals it must be adjusted to the sow's ability to feed the litter, and observation indeed has shown that above a certain optimum number of piglets born the actual number, and not merely the proportion, reared successfully, falls off (see Darlington and Mather, 1949). Given that there is an optimum expression of the character or sub-character, the need for a balance of + and − genes in the system controlling it will follow.

The second essential of the interpretation is that the genic combinations we find are those which have proved themselves under the test of natural selection and hence have survived that test, out of the larger range of combinations that can arise. This has wider implications. Thus the homologous combinations that we find within a population should have been selected for their ability to work together, that is for their relational balance in respect of one another, or to use Dobzhansky's term for their co-adaptation, whereas those of different populations will not have occurred together and hence will not have been co-adjusted by natural selection. Dobzhansky (1948, 1950) has evidence that this is so in respect of the genes carried by chromosome III of *Drosophila pseudoobscura*. If chromosomes differing in, and hence distinguishable by, the inversions they carry are taken from the same wild population and put together in an artificial population in the laboratory, flies heterozygous for the inversions appear to have an advantage in fitness over their fellows homozygous for one or other of the inversions (though not necessarily homozygous for the genes these inversions carry). As a result the population settles down to being polymorphic for the inversions, with each present at its own characteristic frequency, no matter in what proportions the inversions were mixed at the start of the laboratory populations (Fig. 19). The genic combinations of the inversions are evidently co-adapted. If, however, chromosomes marked by the same inversions are taken from different wild popula-

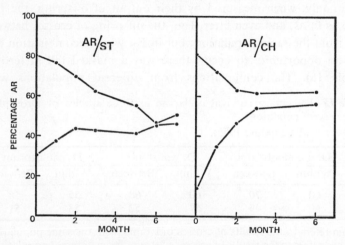

Fig. 19 Co-adaptation (relational balance) in third chromosomes of *Drosophila pseudoobscura* taken from the same wild population. Flies carrying the Arrowhead (AR) and Standard (ST) sequences, on the left, and Arrowhead (AR) and Chiricahua (CH) sequences, on the right, were used to commence laboratory populations. In the case of AR and ST each laboratory population settles down after a few generations to a state where just under 50% of the chromosomes are AR, no matter whether it was started off with an excess or a shortage of AR. The same is true of AR and CH, though here the steady state has about 60% AR chromosomes. In each population a balanced polymorphism has developed because the heterozygotes, AR/ST or AR/CH, is fitter than either homozygote, AR/AR and ST/ST or AR/AR and CH/CH, so displaying the coadaptation of the gene combinations in the two chromosomes AR and ST or AR and CH. (Data from Dobzhansky, 1950).

tions to found the laboratory population, such polymorphisms with their characteristic frequencies of the inverted sequences are not characteristically established: evidently the co-adaptation does not exist. Co-adaptation is thus a feature of chromosomes coming from the same wild population but not of chromosomes from different ones, where natural selection would not have had the opportunity to build it up.

The significance of recombination in the maintenance of relationally balanced combinations is revealed by Vetukhiv's observations (Vetukhiv, 1954; Wallace and Vetukhiv, 1955) on crosses between flies from different wild populations of several species of *Drosophila*. Commonly, when measured by their output of offspring, the F_1's were as fit as, and even fitter than, the offspring of crosses between flies from the same population; but in F_2, when recombination had had an opportunity to occur, there was a marked fall in fertility (Table 15). The combinations from different populations were

Table 15 Percentage survival of larvae in three species of Drosophila under severe conditions.
(Wallace and Vetukhiv, 1955).

	D. pseudoobscura		D. willistoni		D. paulistorum	
	Within	Between	Within	Between	Within	Between
F_1	60	70	60	68	55	58
F_2	60	49	57	52	53	51

Within refers to the results of crosses of strains from the same population. Between refers to the results of crosses of strains from different populations.

reasonably well co-adapted in respect of immediate action of their genes, but were rapidly broken down by recombination into an unbalanced state. Thus the mechanical arrangement of the genes along the chromosomes must be at least as important for the continuing co-adaptation as the adjustment of the numbers of + and − genes to give the immediate balance itself: different arrangements of the genes can give differences in ease of breakdown by recombination even though the overall balance of action is the same (Fig. 20). We may note in passing that in Dobzhansky's experiments, the inversions would serve not merely to mark the chromosomes,

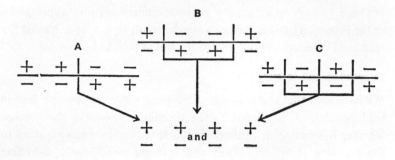

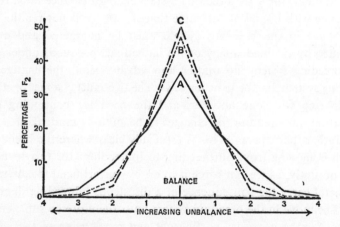

Fig. 20 Linkage and stability of balance in polygenic combinations. All the individual chromosomes have two + and two − alleles, and all three heterozygotes (A – C) are balanced. Because of the linkage organisation of the genes, however, balance is more easily upset by recombination in A than in B, and in B than in C. A single cross-over in A can give the extreme unbalanced types; two are required in B; and three in C, as shown in the upper part of the figure.

The lower part of the figure shows the frequency distribution of individuals with different degrees of unbalance (producable equally by excess of either + or − alleles) in the F_2s from the three heterozygotes, assuming 20% recombination between adjacent loci and no interference. The F_2 from heterozygotes of type A give fewest balanced individuals and the greatest average degree of unbalance. The F_2s from B and C are more alike, but C has somewhat more balanced individuals and a somewhat lower average unbalance.

K

but also to lock together the genic combinations into super-genes in the inversion heterozygotes, with the result that once achieved the relational balance between inversions would tend to be maintained.

Integration and inertia

We have seen that the balance of the genic combinations we find in wild populations is related to the breeding systems of the species, whether inbreeding or outbreeding, and within outbreeders even to the breeding structure within and between populations. Breeding system and balance are intimately related and indeed dependent on each other, for if the breeding system changes balance must tend to change with it, and at the same time if balance is not readjusted the change in the breeding system must be hampered and even prevented by the inadequacy of the individuals produced under the new breeding system, no matter how advantageous the change in breeding system may be in other ways. The two must change together and in step with one another, and the need for continuing co-ordination must require the changes to be suitably gradual, step by step, rather than abrupt. This is especially clear where the breeding system is moving from outbreeding to inbreeding: for this to take place abruptly, as we can force it to do by our artificial means, will in general carry the consequence of a disastrous inbreeding depression. The move in the breeding system must therefore be sufficiently gradual for the necessary sorting out and rebalancing of the genic combinations to go on at the same time.

A chromosome will carry polygenic systems affecting many different characters and the member genes of these different systems must be expected generally to be intermingled with one another along the length of the chromosome. All the systems will be subject to adjustment by selection by virtue of their effects on the characters they mediate, and unless the developmental relations of the characters are close the genic systems controlling them will be prospectively independent of one another, or virtually so, in their adjustments. Mechanically, however, being intermingled along the same chromosome the genes are partially bound to one another, for if a particular combination of the genes of one system is picked out, it will carry with it the combinations of genes in the other systems that are borne in the same chromosome. Equally a cross-over in the chromosome will redistribute the genes of every system, with the result that if a recombinant chromosome is selected for its effect on one character there will be consequential effects on other characters for which no selection is being practised (Fig. 21).

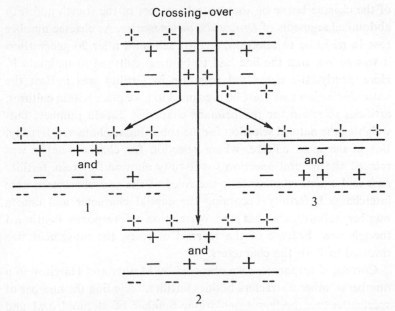

Fig. 21 Correlated responses to selection. Recombination of the genes controlling the primary character (-+-+- and − −) is accompanied by recombination of genes controlling the secondary character for which no selection is practised (+ and −). This may lead to no unbalance (2) or to unbalance in either direction (1 and 3) of the genes controlling the secondary character, according to the position of the crossing-over which gives rise to the recombination. The response of this secondary character to selection for the primary one may thus be in either direction or may be absent. (Reproduced with Professor J. L. Jinks' agreement from K. Mather and J. L. Jinks (1971), *Biometrical Genetics* (2nd edn.), Chapman and Hall, London).

Such correlated responses are a commonplace of selection experiments: successful selection for one character is virtually always accompanied by change in other characters for which no selection is being practised, or even by change in them against the forces of selection to which they are subject. The most commonly observed correlated response is in respect of fertility, which nearly always falls, often disastrously, when selection is practised for other, on the face of it, normally less important characters. To take but one example. Mather and Harrison (1949) selected for increased number of the chaetae borne on the ventral surfaces of the fourth and fifth abdominal segments of *Drosophila melanogaster*. As chaetae number rose in response to selection, fertility fell until after 20 generations it was so low that the line had to be mass-cultured to maintain it. Here clearly, the correlated response in fertility was against the natural selection that must be presumed to take place within cultures, artificial selection for the primary character, chaeta number, outweighing the natural selection for the subordinate character, fertility. But in the mass culture, where selection for chaeta number was relaxed and natural selection for fertility allowed free rein, fertility rose and chaeta number fell: the roles of the two characters were interchanged, fertility becoming the capital character and chaeta number subordinate, but the correlation of response continued though now, because of the reversal of roles, the movement was reversed in both the characters.

Correlated responses were observed by Mather and Harrison in a number of other characters besides fertility, including the number of spermathaecae, body pigmentation, number of sternopleural and coxal chaetae, and mating behaviour. Such correlations would of course arise if the genes that mediated abdominal chaeta number were themselves pleiotropic in action, and affecting these other characters in addition to abdominal chaetae. No doubt such pleiotropically caused correlations do occur at times, but it is unlikely that they are common and most unlikely that they can account for the complex correlations Mather and Harrison found. Indeed even the correlated response in sternopleural chaeta number was not of a type easily explained in terms of pleiotropy and in a later experiment Davies (1971) has been able to rule it out specifically by showing that the genes affecting sternopleural chaetae were distinct from, though linked with, those affecting abdominal chaetae. His results leave no doubt that even with two such apparently similar characters, correlated response springs from the intermingling along the chromosomes of the genes of two separate systems each mediating one of the characters.

Returning to Mather and Harrison's findings, the negative correlation between abdominal chaeta number and fertility, manifested in the first selection and the mass culture, was not final. When selection was resumed from the mass culture, the high chaeta number was quickly re-established and this time without the incubus of poor fertility. Indeed when selection was later relaxed once again chaeta number did not fall, as it did in the first mass culture, even though there was evidence that this second mass culture was not homozygous for the genes controlling chaeta number (Fig. 22).

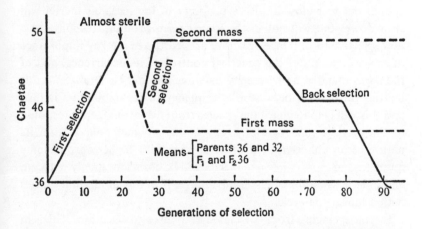

Fig. 22 Selection for increased number of abdominal chaetae in *Drosophila melanogaster* (Mather and Harrison, 1949). In the first selection fertility fell as chaeta number rose. When the selection was relaxed in the first mass, fertility took charge and chaeta number fell. The second selection for chaeta number gave no correlated response in fertility, and the second relaxation of selection was followed by no fall in chaeta number even though successful back selection showed the stock not to be homozygous. Solid lines indicate selection and broken lines mass culture without selection. (Reproduced with Professor J. L. Jinks' agreement from K. Mather and J. L. Jinks (1971), *Biometrical Genetics*, (2nd edn.), Chapman and Hall, London).

Clearly, the genes controlling the two characters had been re-associated by further recombination in the first mass culture, and their balances readjusted to give adequate fertility in company with the high chaeta number favoured by selection.

The significance of these results lies in their showing that although selection for one character can have adverse effects on another, if time is given for the necessary, and it would appear often complex, recombinations to take place between the genes of the systems affecting the two characters, a new level can be established in the one while the other is held at a former and adequate level. In short, the characters are capable of independent adjustment by selection, though because of the mechanical relations of the two systems of genes and need for these to be resolved by recombination, this adjustment may take time. The effect of the linkage is thus not to rigidify the genetic system but to endow it with a property of inertia. This inertia will slow down, but will not in the longer term prohibit, response to a new selective force. Once again we are led to see the key importance of recombination, of the genetical control of its occurrence, and of the adjustment of its frequency and positioning. In *Drosophila* this inertia will be high because the frequency of recombination is low, and this will help to guard the species from disadvantageous response to the cyclical changes of the environment which follow from its nature as an ephemeral. In other species with higher chromosome numbers and more chiasmata inertia will be correspondingly reduced, and indeed would be of lesser advantage, if of any at all, in a species with a longer life-cycle.

Inertia springing from correlated responses to selection, which in turn arises from the intermingling along the chromosomes of genes from unlike systems, is but one aspect of the more general internal adjustment or integration of the genetic system. Reproductive system, breeding system, chromosome system, genic balance and distribution of the polygenic systems within and between chromosomes all affect the flow and release of variability for the many characters of the organism and its response to selection for any one or more of these characters. All have arisen and developed side by side over time and all are adjusted to one another. Change in one will commonly carry with it the requirement for change in another and readjustment of one will imply readjustment of another. An experimenter can maintain his material while these readjustments are taking place successively, as in Mather and Harrison's flies; but in nature, where fitness will depend in a more direct sense on the reaction of the total phenotype, and hence total genotype, to the total environment and its forces of selection, the gains and losses from any change under selection must be kept constantly in closer balance. Adjustment and readjustment of the various characters must therefore proceed side by side in a co-ordinated way, and this must in general imply that change is gradual, going forward only as fast as reassortment of the genes allows the various characters to be kept adequately in step.

Genetic architecture of characters

As we have seen, the polygenic combinations that we find in populations are the ones that have survived the test of natural selection, and they reflect this in the special properties of balance and mechanical organization along the chromosomes that they display, and to which indeed they owe their successful survival. The kind of selection to which they have been subjected and which they have survived also has its consequences, particularly in relation to the ways in which the genes act and interact in producing their effects.

The impact of selection will vary from one character to another, not merely in its intensity but in its nature too. The majority of characters would be expected to be under stabilizing selection, though no doubt there will commonly also be a greater or lesser element of directional selection stemming from a lack of precise agreement between the mean expression of the character and the optimum. This element of directional selection will not however generally be large except in those cases where cataclysmic changes are taking place in the environment. If, on the other hand, we turn to consider fitness itself, it must be under wholly directional selection since *ex hypothesi* it can never be too large. Now fitness is the composite character *par excellence*, since all other characters on which selection is acting will contribute to it. This is not to deny that these other characters may be under preponderantly stabilizing selection, for it is their departures from the optimum that tend to reduce fitness, so that their stabilizing selection will appear as directional selection at the level of fitness itself. Some characters like chaeta number, though making their contributions to fitness, will be less directly concerned with it than will others, like viability, fertility in its various aspects, competitive ability and so on, which may thus be regarded as direct sub-characters of fitness. This does not imply that this second group of characters will not be under some stablizing selection, and indeed we have already seen that litter size in pigs, which must be an important component of fitness, does show evidence of such selection; but it does mean that we should expect them to show more of the impact of the directional selection acting on fitness itself.

Both classes of character show heritable variation, of course, but the structure of the variation is different in the two cases, the so-called 'fitness' characters having a lower proportion of additive variation than the so-called 'peripheral' characters (Robertson, 1955). In a study of chromosome III of *Drosophila melanogaster* Breese and Mather (1960) analysed further the behaviour of on the one hand the genes affecting the number of abdominal chaetae and on the other, the genes affecting viability. The former showed clear evidence of dominance, which was however ambidirectional in that with some genes it was towards lower number and with others towards higher number of chaetae. There was no evidence of non-allelic interaction. With viability, on the other hand, dominance was undirectional, towards greater expression of the character, and there was evidence of duplicate type interaction between non-allelic genes in the direction favouring higher viability. Now, as Breese and Mather pointed out, a gene making for higher viability will commonly be favoured over its allele making for lower viability and we should then expect it to show dominance over the less favoured alternative (Fisher, 1930). With the number of abdominal chaetae, however, where selection is expected to be stabilizing, and intermediate expressions of the character thus favoured, neither the + nor the − alleles would have an unconditional advantage. The commoner allele whether + or − should then show dominance with the average degree of dominance at a lower level (Fisher, *l.c.*). Similarly one might expect non-allelic interaction to develop in one direction and to be of the duplicate type with genes affecting viability, so reducing further the tail of individuals with lower viability, whereas with abdominal chaetae such interaction as did appear would tend to be weaker and ambidirectional, and so self-cancelling.

Thus characters under directional selection would be expected to result in dominance and duplicate type interaction, both pulling in the direction favoured by selection, as Breese and Mather found to be the case with viability, whereas characters under stabilizing selection would be expected to show weaker dominance of an ambidirectional kind and interaction either absent or weak and ambidirectional as has been found with abdominal chaeta number (Mather, 1960; Breese and Mather, 1960). Kearsey and Kojima (1967) have shown that egg hatchability and body weight in *Drosophila* also conform to this expectation, egg hatchability having the same kind of genetic architecture as viability and body weight resembling chaeta number. These authors have further listed a dozen other characters that also conform so far as the evidence goes. Often the weak part of the direct evidence is that relating to the type of natural selection acting on the character, but in the case of sternopleural chaeta number in *Drosophila* there is now ample evidence that it is in fact under the stabilizing selection that its genetic architecture would suggest (Kearsey and Barnes, 1970).

Now in so far as the genes mediating a character which is typically under stabilizing selection show dominance and interaction, even though these are ambidirectional and self-cancelling, they provide the materials for building up new combinations of genes showing directed dominance and interaction if appropriate directional selection is applied. When, in their experiments, Mather and Harrison (1949) applied directional selection to abdominal chaeta number they observed that combinations of genes resulted showing directional interaction. Furthermore these interactions arose in lines selected for both high and low chaeta numbers and they displayed their effects in opposite directions in these high and low lines. Thus not only did selection result in the emergence of genic combinations showing directed interaction from combinations which ostensibly showed no interaction, but the interaction went in either direction according to the direction of the selection applied, as would be expected if the parental combinations had indeed been marked by ambidirectional and hence self-cancelling interactions.

Further evidence has been provided by Spickett and Thoday's (1966) observations on sternopleural chaeta number which they virtually doubled by directional selection. Using Thoday's (1961) special technique to locate the genes chiefly responsible for the change, they found these genes regularly to show dominance in the direction of increased chaeta number and also to show interaction in the same direction, in the sense that two non-allelic genes tending to raise chaeta number jointly produced an effect greater than the sum of the increases for which they were individually responsible (Table 16). Thus here again directional selection has built up directed dominance

Table 16 Interaction of selected genes in their effects on sternopleural chaeta number in Drosophila (Spickett and Thoday, 1966).

0, 1, 2 indicate the number of selected genes present, the alternative gene being from the unselected Oregon stock.

Gene 2 is located on chromosome II and Gene 3a on chromosome III.

Gene 2	Gene 3a		
	0	1	2
	mean numbers of chaetae		
0	20·3	22·8	23·5
1	21·9	26·2	27·2
2	22·1	26·4	27·9

The data give as d values (see Chapter 4) for the two genes $d_2 = 1·55$ $d_{3a} = 2·25$ and $i = 0·65$ where i is the interaction of the two d's (see Mather and Jinks, 1971) and is statistically significant here.

The data also display dominance of both selected genes over their Oregon alleles.

and interaction. True, the interactions were of the complementary kind, the two genes when together having a disproportionate effect in the direction of selection rather than of the duplicate type with a disproportionate effect away from the direction of selection as observed for viability by Breese and Mather. This, however, is not to be regarded as discrepant since complementary interaction is to be expected when selection has been recent, heavy and with emphasis on the attainment of a high expression of the character by a few individuals chosen as parents. Directional selection of long standing, emphasizing the achievement of a high proportion of individuals all of roughly equal merit such as must appertain in nature, would give duplicate interaction which maximizes the proportion of individuals with high adaptation at the expense of the tail of the poorly adapted. In other words the difference in the type of interaction yielded by Spickett and Thoday's selection on the one hand and natural selection on the other precisely reflects the difference in the types of the selection themselves. With suitable continuation and modification we might expect the artificial selection to produce the same duplicate type of interaction as does natural selection.

Like the breeding system and balance, the genetical architecture of characters reflects the past action of selection and in doing so it must in some measure govern the organism's response to the future impact of selection, though the rapid changes observed by Mather and Harrison and by Spickett and Thoday suggest that adjustment of genetical architecture by selection is not subject to the same degree of inertia as is adjustment of breeding system and balance. They serve, however, to emphasize that the attainment of the architecture, or indeed the balance, that selection would be expected to favour is not limited in the long run, or even seriously hampered, by any restriction arising from the innate properties of variation of the genetic units themselves. The readjustments can be achieved by reassortment and reassociation of genes already present in the gene pool and within wide limits readjustment will proceed as far as the particular force of selection will push it.

8   The Consequences of Disruptive Selection

Nature of disruptive selection

So far we have concentrated attention on selection towards a single optimum phenotype, though we have brought into the discussion consideration of various types of change in this single optimum. In general, however, because of spatial variation in the habitat of a population and functional variation among the individuals of a population different individuals must be subject to different pressures of selection: there must prospectively be more than one optimum phenotype. In such a case, as we saw briefly in Chapter 5, selection is said to be disruptive (Fig. 9). We must now turn to consider the consequences of such disruptive selection and we shall take the opportunity to examine further the effects of selection where there is a single optimum at any one time, but an optimum which changes cyclically over the generations. For the sake of simplicity we will concentrate chiefly on the case of two optima, as no new principles are involved when the number of optima is greater than this.

Where more than one optimal phenotype is favoured simultaneously in the population and these optima persist over the generations there must be a tendency for selection to produce a corresponding number of groups of individuals within the population, each group being selected towards one of the optima. The extent to which two or more distinguishable groups are actually produced and the genetical relations between such groups as may arise will depend on the circumstances prevailing within the population, including:

1. The pattern of selection and in particular the intensity of the

selective forces favouring each optimum phenotype and disfavouring intermediate phenotypes.

2. The rate at which genes are exchanged between the prospective groups, each under selection towards a different optimum.

3. The relations between the optimal phenotypes, whether they are independent of, or dependent on, each other for their maintenance or functioning.

There is now a substantial body of experimental evidence on the action of disruptive selection, chiefly from Thoday and his collaborators to which we may turn for information (see Thoday's valuable review, 1972).

The differences in selection pressure favouring the different optimal phenotypes, may arise from separation of the individuals in space, as where differences within the population's habitat impose on the individuals a need for adjustment to different 'niches' within the environment; from separation in time, as where successive generations, or even different sets of individuals within a generation, meet different conditions because they are not developing and reproducing simultaneously; or from separation in neither space nor time but by the functioning of an individual depending on other individuals acting in a different though related way.

Polymorphism

Taking this last situation first, separation by function is perhaps seen at its most obvious where, as in most animal species and many plants, there are two different classes of individual, males and females, discharging different but complementary functions in reproduction. There are thus two optimal phenotypes, one for each sex, which though different must be closely adjusted to one another because they are bound together by the function which they jointly discharge or, to put it another way, whose labour they divide between them.

Gene flow between the two phenotypic groups, males and females, is not merely free, it is enforced since, sex chromosomes apart, every individual no matter to which group it belongs derives half its genes from a parent of the opposite group, as well as half from a parent of its own group. The X chromosome, where one exists, is also common to the two groups, and only the Y which as we have seen is often virtually inert or may be absent, is confined to one group. The disruptive selection acting on the two phenotypes cannot therefore result in any tendency for the general genotype of the population to break up into two divergent types. Nor can a compromise common genotype producing an intermediate phenotype be favoured, since intermediates between the two sexes are sterile in most animals and in any case if capable of functioning, as are hemaphrodites in dioecious plants (see p. 110), will jeopardize the outbreeding system which sex separation provides. The common genotype must therefore be such as to offer two, and only two, closely regulated and adjusted channels of development, one leading to effective males and the other to effective females. Thus the first effect of disruptive selection to note where the optimal phenotypes are bound together is the production of a common genotype which offers an appropriate number of closely canalized but divergent lines of development.

Given, however, that the two (or more) channels of development are available, something must switch the individual's development into one or other of them. Most commonly this switch is genetic, depending on the segregation of an effectively unitary genetic difference in one or other sex (see Chapter 6). It need not however be so for the line of development leading to male or female, may depend on the environment in which the zygote finds itself, as we have seen in the marine worm *Bonellia*. In such a case the common genotype must be adjusted by selection not only to offer the two channels of regulated development, but to move into one or other according to the appropriate feature in the environment that is encountered.

These principles are not confined in their application to sexual dimorphism. They will obviously apply to any outbreeding system depending for its functioning on diversity as do heterostyly or other forms of incompatibility, and as we saw in Chapter 6, these breeding systems have the same genetical structure, comprising an appropriately adjusted common or background genotype and a switching mechanism depending on the segregation of effectively unitary genetic differences, as do most cases of sex separation. The principles apply, however, more widely still, wherever in fact two or more different phenotypes depend on one another for their functioning. Thus in Batesian mimicry among insects, the mimic which is liable to predation secures protection by mimicking a model which is distasteful to predators. But since the predators must learn to associate with distastefulness the appearance shared by model and mimic, the mimic will lose its advantage if it becomes over-common relative to the model. Thus there is advantage in the mimic species mimicking not one but a number of models, with sufficiently small numbers of mimic individuals resembling each of the various model species. There can even be advantage in some individuals of the mimic species mimicking no model at all if this leads to an increase in the prospects of them all. The various phenotypes are thus interdependent because the differences among them, arising from their separate adjustments to different optima, are to the advantage of them all. We would therefore expect a polymorphism with the same genetic structure as in sexual dimorphism – a common genotype adjusted to offer alternative channels of development towards the various morphs, and a switching system dependent on simple genetic segregations. The expectation of polymorphism is commonly realized in species showing Batesian mimicry, and in *Papilio dardanus*, where they have been tested, the detailed genetic expectations are also borne out (Clarke and Sheppard, 1960, 1962). The significance of the disruptive selection is emphasized by comparisons of species showing Batesian mimicry with others showing Mullerian mimicry, in which distasteful species mimic one another, so achieving economy in the education of predators to leave their joint phenotype alone. There is thus no disruptive selection in this type of mimicry and, as might then be expected, species having Mullerian mimicry do not characteristically display polymorphism.

In Batesian mimicry the switching mechanism in genetic and indeed it is difficult to see how an environmental switch could operate. In other cases, however, the switch is environmental. The female honey bees of a hive are at least dimorphic, comprising the queen, who reproduces, and workers who maintain the hive, forage for food and care for the young. The development of a larva into one or the other depends on the food given to it by the workers, that into a queen being determined by the 'queen substance'. The flexibility that such a switch gives to the development and reproduction of the hive as a whole requires no comment, and we need only observe that the behaviour of the workers in applying the nutritional switch to the larvae in appropriate circumstances must itself reflect appropriate adjustment of the common genotype. Ants offer examples of more complex systems having no doubt a basically similar type and determination.

The quasi-polymorphism of a hive of bees or colony of ants rests on differences among the individual insects, though it is internal to the hive or colony which is the unit of over-all functioning and particularly of survival and reproduction. We see, however, the same principles at work within individuals. The differentiation among the cells, and the tissues into which they are aggregated, is a kind of polymorphism not now of individuals but of their parts, which differ according to the functions they divide amongst themselves, yet must be mutually adjusted to work together and are bound together in survival and perpetuation of the system. These cells share a common genotype adjusted to offer the development of appropriate and mutually compatible cell types. If this genotype is broken down, so too is proper development and differentiation as we can see, for example, in wide echinoderm crosses like that between *Dendraster* and *Strongylocentrotus*, where early development follows the normal pattern of the mother species so long as the maternal cytoplasm is in charge but becomes mixed at the gastrula stage when the unadjusted hybrid genotype begins to take control (Waddington 1939).

The switching system in differentiation obviously cannot depend on gene segregation, and must depend on interaction of the genes with their cytoplasmic surroundings (Mather, 1948). The way in which the genes play their parts has been shown by Jacob and Monod (1963), who elucidated the complex interactions of repressor, operator and structural genes involved in the induction of *Esherichia coli*; but in the system they analysed the initial switch was external in that it depended on the presence or absence of chemical substances in the environment. In differentiation switching obviously cannot be environmental any more than it can be genetic. It must be internal and must depend on some form of auto-triggering acting via the cytoplasm and depending on earlier interactions of the genes with the cytoplasm. Whatever the precise nature of this switching, we see that at the cellular level as well as the level of individual the same principles apply. Division or specialization of function requires the production of types, whether of cells or individuals, each different and individually closely adjusted to its role but compatible with the others, and therefore all bound together in their operation. The task of providing the necessary, closely-regulated, paths of development falls on the common genotype, which is adjusted by selection to discharge it, while the switching of cell or individual into one or other of these paths may also be genetic, resulting in a segregating switching gene or super-gene, may be dependent on the external environment, with the common genotype adjusted to respond to or even determine the environmental difference, or may be dependent on the progressive interplay of genes and cytoplasm as in differentiation (Table 17).

Table 17 Examples of various forms of switch mechanism in polymorphisms and quasi-polymorphisms.

In all cases the common genotype must be adjusted to offer the alternative paths into which development can be switched, and to endow the system with the capacity for being switched.

	Switch	Polymorphism or Quasi-polymorphism
External	Spatial	Sex-determination in *Bonellia*. Phase in migratory locusts.
	Seasonal	Reproduction in aphids.
	Nutritional	Induction in micro-organisms. Caste in honey-bees.
Internal	Nuclear genes	Sex-determination in most cases. Incompatibility in plants. Most polymorphisms. (e.g. mimicry, blood groups, etc.)
	Fertilization	Sex determination in Hymenoptera.
	Cytoplasm	Differentiation.

L

The action of disruptive selection in building up both the common genotype and the genetical switch that we commonly see in polymorphic populations, is well shown by the experiments, especially those of himself and his collaborators, reviewed by Thoday (1972), and in particular by one which involved parallel selection in *Drosophila melanogaster* for high and low numbers of sternopleural chaetae with mating between selected high and selected low flies. The experimental situation thus corresponds closely to sexual dimorphism or heterostyly, with high and low chaeta flies corresponding to the two sexes or to the heterostyled pins and thrums. Polymorphism was built up, with the population containing two distinct types having high chaeta numbers and low chaeta numbers respectively. A switching system came into being on chromosome III and the highs and

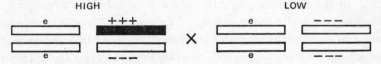

Fig. 23 Polymorphism for sternopleural chaeta number built up by disruptive selection in *Drosophila melanogaster*. The third chromosome provides the switch, which has been built up out of genes at three loci, + indicating an increasing and − a decreasing allele. The second chromosome, carrying ebony (e), enhances the difference in chaeta number switched by the third chromosomes. There were probably other enhancers to the first chromosome. (Reproduced by permission of Professor J. M. Thoday and The Royal Society from J. M. Thoday (1972), 'Disruptive selection', *Proc. Roy. Soc. B.*, **182**, 109 – 143.)

lows differed in respect of this chromosome, with the lows homozygous for the low chromosome (i.e. LL) and the highs heterozygous for it (i.e. HL). Genes were accumulated in chromosome II, and probably chromosome I also, which enhanced the innate difference in effect on chaeta number of the H and L forms of chromosomes III. Chromosomes I and II which were uniform in the population and common to both high and low morphs thus provide the common background which endows the individuals with the capacity for being either a high or a low, and the H and L chromosomes III provided the switch as do the sex chromosomes or the S–s difference in heterostyly. (Fig. 23).

Thoday's switching difference was itself compound, consisting of a linked complex of three loci such that an H chromosome carried the + alleles and the L chromosome the − alleles at all three loci. In this too it not only revealed the way that selection has built it up from its constituent parts, the individual alleles, but also resembled the compound switching difference that is so clearly seen for example in the heterostyled Primulas (Chapter 6).

The aggregation of the three loci into a super-gene was not, however, complete as in so far as recombination could and did continue to occur between them in the HL heterozygote of the high morph, the switching complex was continually tending to break down. Should an inversion arise embracing these three loci, this breakdown would cease, or at least be materially reduced, since HL would be heterozygous for the inversion as well as for the + + + and − − − aggregation of alleles. Fisher (1930) and others have argued that in such a linked complex natural selection should be tending to reduce the frequency of recombination and hence of breakdown, but the design of Thoday's experiment was such that it could give no evidence on this point. In any case, in another sense, the occurrence of recombinational breakdown in Thoday's switching complex serves to emphasize still further its similarity to, for example, the S–s complex in heterostyled Primulas where at times breakdown still occurs by reassortment of the constituent parts, presumably through recombination.

The effects of breakdown of such a switching super-gene S–s in distylic Primulas serves to remind us in a further way of the two different genetical adjustments involved in the rise of a genetically switched polymorphism. The observed recombinants show reassociation of the stamen and style types, so that anthers and stigma are borne at the same level to give either long or short homostyles (see Fig. 24). The incompatibility properties however, have not been altered, with the consequence that the homostyles are as capable of self-pollination as pins and thrums are of crossing. In addition, of course, a long homostyle will be fully effective as male-parent but incompatible as female in crosses with pins, and effective as female parent but incompatible as male in crosses with thrums, the relations being reversed with short homostyles. In other words, the incompat-

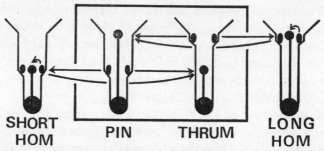

SHORT
HOM   PIN   THRUM   LONG
HOM

Fig. 24 The breakdown of distyly in Primula by recombination within the S gene, reassociating anther and stigma positions. Where the reassociation gives the pin position for anthers with the thrum position for stigma, a short homostyle results (left), and thrum position for anthers with pin position for stigma gives a long homostyle (right). The arrows show compatible pollinations, which include selfing in both homostyles. The normal outbreeding, distylic system of pin and thrum is enclosed in the box.

ibility system is still fully operative but now determines a different set of successful combinations in pollination, not all of which are crosses. When, however, the balance of the common genotype has been broken down, as it has by inbreeding and selection in *Primula sinensis* (Darlington and Mather, 1949), the manifestation of heterostyly becomes irregular in both its morphological and its physiological manifestation, with the result that while, as with breakdowns of S–s, both selfing and crossing can go on, they do so in a less regulated and more haphazard way, and not in new but equally precisely determined combinations as when the S–s alleles are altered. Breakdown of S–s results in the system operating in a new but still precisely regulated way: breakdown of the common genotype impairs the precision of functioning of the system and tends to destroy the system as a whole. With more ancient switching systems, like many of those involved in sex-determination, the clear distinction between switch and common genotype may be lost, as can be seen in cases of sex-determination where numerical balance between X chromosome and autosomes acts as the switch. In such a case any breakdown is liable to affect the functioning of both switch and common genotype. The result is thus an ineffective, and in sexual dimorphism sterile, intermediate type.

Just as switching super-genes have a complexity which is revealed by breakdown we must suppose that they can grow and become more complex as the morphs acquire new and newly favoured differences between themselves. There is insufficient reliable evidence to provide an adequate basis for further discussion of this possibility, but should a new function be added to a polymorphism and the super-gene grow accordingly, there might well come a time when the older functions were no longer critical, with the newer function becoming the one by reason of which the polymorphism was maintained. The significance of the older parts of the switching system would then disappear and their presence could only be understood by reference to a past no longer easily traceable. Thus if polymorphism is a phenomenon which grows, it will inevitably have its own archaeology, similar to that we have seen in a different context where inbreeding mechanisms are super-imposed on cross-breeding (see p. 116).

Developmental plasticity

Changes in the environment over time lead to individuals of different generations (and possibly even of the same generation) encountering different environments and therefore different selective forces. Where such changes recur, and in particular where they are cyclical, different optimal phenotypes will be favoured in a recurring sequence, even though the same general optimum may apply to all the individuals co-existing in the population at any one time. There will thus be a form of disruptive selection acting over time on each lineage, and hence on the population as a whole, even though selection is towards a single optimum at a given time.

Certain consequences of such selection have been discussed earlier when we were considering selection towards a single optimum. The situation is frequently met by the production of a phenotype which, if fully adjusted to only some of the variant environments, is sufficiently satisfactory in all of them and hence affords an acceptable compromise over the environmental cycle. The disadvantage of erosion of variability by fruitless response to cyclically changing forces of selection can then be kept within acceptable bounds by inbreeding or restriction of recombination. A different response to this disruptive selection over time is however possible, and one that has close similarities with polymorphism.

A basic property of polymorphism is, as we have seen, the common genotype adjusted by selection to offer each zygote the appropriate range of distinct, well-regulated channels of development. Which channel will be followed depends on the switch which, though commonly genetic, is at times environmental as with sex in *Bonellia*. Now the operative environmental difference with *Bonellia* is between environments existing simultaneously; if the fertilized egg falls free in the sea it becomes female, and if it falls onto a pre-existing female it becomes male. Should the environments which switch development not co-exist but occur at different times, however, the result would be not a polymorphism of the familiar type, but a kind of temporal or sequential polymorphism. Seasonal differences could clearly provide such a switching system and cases of it are known. The most familiar, which we have already had occasion to note in a different connection, is aphids which reproduce sexually in the winter, but where summer generations reproduce by pathenogenesis, no doubt thereby securing the advantage that where numbers are rising dramatically the increase in variation which under sexual reproduction accompanies such a rise is avoided (see Darlington and Mather, 1949). The continuing lineage of the aphids depends on successfully coping with both winter and summer conditions as they successively arise, and the two phenotypes, sexual and parthenogenetic, are thus bound together by mutual dependence in descent, just as the morphs of the more familiar polymorphism can be bound together by interdependence in contemporary functioning.

Such behaviour in the aphids may also be regarded as a form of developmental plasticity adjusted to seasonal changes. When looked at in this way, both it and for that matter environmentally switched polymorphism have much in common with more familiar cases of plasticity such as are shown for example by plants which may find themselves growing in running water or aerially on land, albeit wet land, and which respond by having differently structured leaves (e.g. *Ranunculus aquatilis*). They differ from such cases of plasticity in one important respect: in the aphids and *Bonellia* there is a fixed bond between the two optimal phenotypes, in the one case a mutual dependence in descent and in the other a mutual dependence in reproduction. There is no such fixed bond in the water crowfoot: they do not need to breed together, nor are they bound to be joined in maintaining the lineage. They could, at any rate, in principle therefore become genetically isolated from one another in the way to be examined in a later section, and so lose the common genotype and the need for plasticity, which could not happen without the whole way of life changing where the phenotypes are bound together. Whether this will happen in a case such as the water crowfoot will depend on various things – for example, on the occurrence of unusually dry seasons when a regularly aquatic plant would be at a grave disadvantage, which is tantamount to saying on the occurrence of seasons when one form would be dependent on the other for maintaining its descent. In any case once having achieved a genotype mediating suitable plasticity there may be little advantage in any break-up of the population.

One last point remains to be made about temporal or cyclical polymorphism. Since at any given time the same selection pressures will be impinging on all the individuals, at least on all those in a given place, this type of selection, disruptive only over the lineage in time, cannot of itself favour genetical diversity among the individuals of the population. In this it stands in sharp contrast to the effects of disruptive selection produced by separation in space, to which we must now turn.

Divergence

There can be few, if any, environments which are uniform over the whole range occupied by a population. The differences may be small or large relative to the fluctuations in the environmental condition over time and in particular from one year to another; but even if relatively small they will introduce some local bias into the environmental conditions and the different individuals of the population will to this extent be subjected to varying forces of selection. Assuming that there is free migration, or to put it another way free gene flow, between the different localities, this variation in the impact of selection must generally reduce the extent to which one gene is favoured over its allele since this advantage will fluctuate, and may even change direction, from one locality to another. We would therefore expect local variation of the environment to result in an increase in genetic variation within the population as a whole, and this expectation is borne out by experience from a wide range of experiments (Thoday, 1972).

This increase in variation was shown in a most interesting way by the experiment, quoted by Thoday, in which Powell (1971) set up populations of *Drosophila willistoni* in laboratory cages, some with but one kind of medium, with one kind of yeast and kept at a constant temperature. Other cages had more than one kind of medium or more than one kind of yeast or were subjected to weekly changes of temperature, while still others combined all these variations of the environment. Though starting with flies from the same stock, the different cages showed differences a year later when tested for variation at twenty-two loci mediating the production of enzymes. Individuals from the populations with uniform environments were on average heterozygous at 7·8% of the loci; those from environments with one variable averaged 9·6% heterozygosity; and those from environments with all three variables showed 13·4% heterozygosity. Clearly greater variation was maintained in the more variable environments, and the data even suggest that the increase in variation is approximately proportionate to the number of variable components in the environment, though further observations would be required to substantiate this.

Where there was a gradation by small differences between localities, the environment, and with it the selective forces and hence the optimal phenotypes, would be showing continuous variation. Characters under polygenic control in a population inhabiting this range of localities would be expected to show greater phenotypic and genotypic variation than in a more uniform set of localities, but no striking differences or divergencies would be expected to arise. Where, however, there were a few localities, or groups of localities, with larger selective differences between them, and hence delimiting only a few optimal phenotypes with larger selective pressures towards them, the situation would approach that which we have already described in relation to differential functioning between individuals and we should expect a much greater tendency for the population to develop groups with greater phenotypic and genetic differences between them. This situation has been mimicked by Thoday and his collaborators in experiments in which disruptive selection has been imposed for high and low sternopleural chaetae in *Drosophila melanogaster*. In these experiments the two directions of selection represent the disruptive selection imposed by two different localities or habitats, and gene flow such as would follow migration between localities is represented by appropriate matings between individuals selected in the two directions, or descended from parents selected in the two directions. The amount of gene flow between the two groups, selected towards high and low chaeta number respectively, was regulated by the frequency of their cross-matings.

The results of these experiments are listed by Thoday (1972), who also gives references to the more detailed accounts to be found in the original publications. It will suffice for our present purpose to note that even with 50% gene flow (that is with each group of flies, H and L, deriving half their genes in each generation from the other group) divergence occurred and was maintained between the average phenotypes of the two groups after about a dozen generations of selection. With 25% gene flow (which is the equivalent of random mating in an experiment comprising two groups of equal size) consistent divergence between the phenotypes of the groups was obtained more quickly and the divergence was greater than with 50% gene flow: indeed in the long run divergence with 25% gene flow was as great as with no gene flow at all. Thus divergence occurs even where the gene flow exceeds that which would be given by random mating, and at the random mating level of flow divergence can be very substantial.

We may observe that 50% gene flow is achieved by ensuring that all matings are between individuals from the two different groups, which is of course tying the two groups together in the reproductive function. It is not surprising therefore that, as we have already seen to be the case in a different experiment, a polymorphism was the result. The system of selection in the two experiments was not, however, quite the same. In that referred to in the earlier section, low flies were taken from the L group, and mated to high flies from the H group: the disruptive selection was complete and a simple polymorphism resulted. In the experiment to which we are now referring, on the other hand, high flies were taken from the L group to mate with high flies from the H group, and low flies from H to mate with low flies of L. The analogy with sexual dimorphism is thus incomplete in this experiment, which indeed more nearly resembles local selection in two habitats with selected migration from one to the other, as Thoday observes. It is not surprising therefore to find that although a polymorphism resulted, its genetic structure was not of the typical kind, in that its switching system was not only composite, giving three morphs one of which was the 'migratory' morph, but was also much more confounded with the background genotype. The different selective relations of the groups in the two experiments appears to have resulted in a different genetical architecture of the polymorphism.

The finding that even 50% gene flow does not prevent genetical divergence of the groups is a dramatic demonstration of the power of disruptive selection. At the same time it should be recognized that in the experiments of Thoday and his collaborators, the selection exercised was very powerful, only one fly being selected out of 20 as a parent for the next generation. Less powerful selection was exercised by Streams and Pimental (1961) in a somewhat different type of experiment again using sternopleural chaetae in *Drosophila*. Rather than practising gene flow between two groups under simultaneously selection in opposite directions, they fed into their selection lines flies taken at random from the base population from which all the selection lines were derived. With a selection pressure of one fly selected out of 10, for use as a parent in the next generation (i.e. half Thoday *et al's* intensity of selection), 50% gene flow gave but little, if any, divergence of the selected line from the base population.

though 20 % gene flow permitted considerable divergence. When the selection pressure was reduced to 40 % (i.e. two flies out of five or four out of 10), some divergence was still obtained though there are some doubts about its precise significance. Be this as it may, however, in conjunction with those of Thoday *et al*, this experiment demonstrates the simultaneous dependence of divergence on the pressure of the disruptive selection and the rate of the gene flow between the groups: the greater the pressure the greater the flow at which divergence can be obtained (Table 18). It is not surprising therefore that

Table 18 Selection pressure, gene flow and divergence under disruptive selection.
(Results from Thoday *et al*. and Streams and Pimental, quoted by Thoday, 1972).

	Percentage of flies selected for breeding (less ⟵—selection pressure—⟶ greater)		
Gene flow	40 – 50%	10%	5%
50%	—	?	D
20 – 25%	d	D	D
0% (isolation)	D	D	D

D = marked divergence.
d = slight divergence.
? = doubtful divergence.

examples of divergence over short distances in natural populations of plants have involved heavy selection pressures, as in the example where Bradshaw and his colleagues (see Bradshaw, 1971) have demonstrated the rise of genetically divergent types of plant in extreme habitats even where these were so close to more normal habitats (the separation sometimes being no more than a few metres) that migration of both pollen and seed would ensure a high rate of gene flow.

Isolation

When divergence takes place as a result of disruptive selection in two localities, crosses between the two divergent groups will result in individuals which are genetically adjusted to neither locality. Unless there are still further localities into which they might fit, these individuals will be at a disadvantage either immediately, if their phenotypes are intermediate between those of the divergent groups, or in the next generation, when segregation occurs, where as a result of dominance they resemble in phenotype one or other of the divergent groups. Thus with divergent groups, there will commonly be a penalty to be paid, whether immediate or deferred, by individuals resulting from crosses between them. Any gene which tends by whatever means to reduce crossing between the groups will thus confer an advantage on individuals carrying it, as they will thereby be rendered less liable to produce offspring with reduced fitness. In principle, therefore, we would expect divergence to stimulate the rise of genetic mechanisms reducing the frequency of inter-group crossing; except of course where the groups which the disruptive selection has produced are functionally bound together as males and females or incompatibility groups are bound in the reproductive function itself.

The selective advantage of such bars to crossing, and the selective pressure towards their evolution, will depend on the degree of disadvantage of offspring resulting from inter-group crosses. The advantage may be reduced or even absent where there are localities which favour intermediate phenotypes and which are readily available to the offspring of crosses. In other words continuous gradation in the optima favoured by a range of readily available habitats will reduce the advantage of bars to crossing. The fewer and more sharply distinguished the optima favoured in different places, the stronger will be the move towards reduction in crossing, for the greater the disadvantage of the crossed individuals and their segregating offspring. Equally, the greater the gene flow is between the groups whether from migration or other cause, not only will divergence be correspondingly more hampered, but the disadvantage of cross-bred individuals will also be less since a greater proportion of offspring will be from crossing and the competitive disadvantage experienced by any one of them will thus be reduced. At the same time any divergence will mean some disadvantage of crossed offspring which will tend to stimulate bars to crossing, and in so far as any reduction in crossing ensues the greater will be the tendency to divergence. Any consequent increase in divergence will result in increased disadvantage of crossing, which in turn will then put an increased premium on genes reducing the rate of crossing. Once started therefore the system becomes self-stimulating in its development: hybrid incapacity stimulates bars to crossing which in turn stimulate further reduction in crossing and hence increase in divergence until eventually gene flow effectively ceases between the groups which thus become genetically isolated from one another (Mather and Edwardes, 1943).

This self-stimulating progress may be set off in more than one way. Where disruptive selection produces divergence, the relative incapacity of the inter-group crosses would appear to be the starting point. Under other circumstances it could be a reduction in crossing. We have seen in Chapter 7 that whereas polygenic combinations from the same population are mutually balanced or coadapted, and hence both work harmoniously with one another in producing fit offspring and maintain themselves satisfactorily over the generations, similar combinations borne by the same segments of chromosome, but coming from geographically different populations do not do so. When brought together in heterozygotes they may give less fit individuals and they tend to break down by recombination in F_2 and later generations. They have lost both their relational balance of action and the mechanical relation that enables them to maintain their integrity sufficiently well when they meet. Such a result could be ascribed to the action of disruptive selection, but it would not be disruptive selection as we have been using this term, for it would be towards different optima which characterized different populations, not towards different optima occurring within a single population.

Populations that are geographically separated will in general have very little if any gene exchange. With virtually zero gene flow it would take little difference in the impact of selection on any character of the two populations to result in the rise of a difference of polygenic balance much greater than the phenotypic difference with which it was associated. Any small genetic difference which selection produced would in turn be steadily magnified by the stochastic processes involving recombination and balancing that are inevitable in populations of finite size; and indeed, as we have seen, the phenomenon requiring explanation is the coadaptational properties of polygenic combinations within a population rather than the lack of them in combinations from different populations. Even in the absence of difference in the selective forces impinging on separate populations, the normal flux of polygenic variability is likely to result in small divergences arising in the polygenic combinations mediating the character, and since every change would favour consequent adjustments at other loci these combinations would then tend to move further away from one another along divergent paths of successive genetical adjustments, even though all towards the same phenotypic expression.

Thus the prevention of crossing by geographic separation of populations would be expected to lead to differences in genic adjustment which would be reflected in reduced fitness of hybrids between them. This in turn would favour the rise of bars to crossing, that is the rise of genically determined isolating mechanisms, should the populations or their descendants later come into geographical contact with one another. Where this is the course of events, the genic isolating mechanisms should be more in evidence in contiguous or overlapping populations, and there is evidence that this is so. And should human intervention or cataclysmic change in the environment result in hitherto non-contiguous populations being brought into contact, hybridization could occur, though the hybrids would in general be less fit and so without effect on the parent populations over the generations, except where by crossing back to one parent type or the other they resulted in introgression of one type by genes from the other (Anderson, 1949).

When the genic bars to crossing, arising in this way, have become strong enough effectively to prevent any gene flow between the groups, apart from occasional introgression, the populations are effectively isolated genetically and will therefore go their own genetic ways in future. They have in fact become separate species which have arisen because of the initial geographic separation and are hence said to be of allopatric origin. It has been argued, notably by Mayr (1954) that such allopatric speciation is by far the chief, even if not the only, way by which separation of one species from another comes about. Given, however, that despite gene flow disruptive selection can produce divergent groups within populations, with the consequences that the progeny of inter-group crosses will often be at a disadvantage and that the rise of genic bars to crossing will be favoured, isolation should be capable of coming about within the range of a population. Speciation could then take place sympatrically.

We may note in passing that groups within a population may be isolated from one another by means other than sexual. In the fungus, *Aspergillus nidulans*, cytoplasmically determined disabilities may be passed from one individual to another by the fusion of their hyphae to produce heterokaryons (Caten, 1972). Such heterokaryons bear nuclei of more than one genetic type and these nuclei must have genotypes capable of working successfully together and of surviving recombination with one another when the heterokaryon undergoes sexual reproduction. The frequency of hyphal fusion is restricted, and with it, as Caten has shown, the spread of cytoplasmic infection, by the occurrence within populations of groups such that although individuals within a group form heterokaryons readily under appropriate circumstances, individuals from different groups cannot do so. This heterokaryon incompatibility is genetically determined, and as with sexual or geographical isolation, the incompatibility groups appear to have lost the co-adaptation of their genotypes (Jinks, *et al*, 1966). Indeed these heterokaryon incompatibility groups in some respects are like a swarm of sibling species, resembling one another phenotypically but separated from one another genetically just like certain groups of species in *Drosophila*. This swarm must, however, have arisen sympatrically as it is only when individuals are in contact with one another that the system of heterokaryon incompatibility can confer an advantage through the limitation it imposes on the spread of cytoplasmic infection.

Returning to our main theme, there is experimental evidence of genetic isolation arising sympatrically under the action of strong disruptive selection. Thoday and Gibson (see Thoday, 1972) practised selection for high and low numbers of sternopleural chaetae in *Drosophila melanogaster* in experiments where the selected flies of both kinds were placed together in a single vial for mating. The flies then had a choice of mates: they were free to mate at random, or for H to tend to mate with H and L with L, or for that matter to mate preferentially in a disassortative way, H with L. After only 12 generations in one such experiment and no more than seven in another, the distributions of chaeta number in the progenies of H and L females no longer overlapped and it was possible to show that this was because hybrid flies were just not produced by crosses between H and L parents. The rise of divergence between the H and L groups had been accompanied, indeed facilitated, by the rise of some presumably genic bar to crossing between them. The initial disadvantage of flies with intermediate numbers of chaetae from crosses between selected H and L parents had put a premium on the avoidance of such cross-matings and the self-stimulating progress of divergence and reduction of crossing had been set in motion, the divergence becoming greater and the reduction in cross-mating presumably stronger as the number of selected generations rose. Furthermore, Thoday and Gibson were able to show by direct experiment that the reduction in cross-mating was due at least in part to the exercise of choice in mating, the ratio of $H \times L$ cross-matings to $H \times H$ and $L \times L$ intra-group matings being $\frac{11}{78}$ and $\frac{42}{133}$ respectively in their two experiments.

Isolation has not been found to result in all such experiments, but the reasons for the different outcomes of different experiments are not clear. It is, however, clear that at least some populations have the capacity for building up isolating mechanisms under the action of disruptive selection and doing so remarkably quickly. Isolation, with its corollary of sympatric speciation, is thus a prospective outcome of disruptive selection, when the optimal phenotypes favoured by the selection are not bound together in the achievement of fitness, just as polymorphism is the outcome when the phenotypes are so bound.

Isolating mechanisms

Gene flow between two groups of individuals can be reduced and isolation of the groups from one another achieved, in the sense of restricting gene flow to the minimum at which the groups go their separate genetical ways, by either or both of two means. In the first place, hybrids between the groups may be unfit because they are developmentally inadequate, sterile or ill-adapted to survival and effective breeding in the prevailing environmental circumstances. The hybrids then fail to leave progeny or at least to leave progeny on the scale achieved by parental groups. The second means is by the action of bars to crossing between individuals of the two groups and the consequent failure of hybrids to be produced. Though both means will prevent gene-flow, they must come about in different ways.

The production by an individual of incapable hybrid offspring as a result of crossing with another individual from a different group, cannot in general result in other than a reduction in the prospects of that individual leaving descendents in the second and later generations. In other words, the production of incapable hybrid offspring represents a lowering of parental fitness. There can thus be only selective disadvantage in hybrid incapacity *per se*, and it cannot arise therefore by the action of any selective force directly favouring it. On the contrary, its occurrence is simply a passive consequence of the genetical divergence consequent on the two groups being adjusted by disruptive selection each to its own special circumstances despite gene flow between them, or on the divergence between two groups which though capable of crossing are prevented by geographical separation from doing so and so go their separate genetical ways.

Bars to crossing are in a different case. Any gene which reduces the frequency of crosses giving rise to incapable hybrids will confer an advantage on individuals carrying it provided that the reproductive effort so saved from inter-group crossing is available for intra-group matings yielding fit offspring. Unlike hybrid incapacity therefore, genetical bars to crossing will be built up by the direct action of selection. They will come about, however, only in so far as the hybrids whose production they are preventing would be inca-

M

pable, whether as the reverse side of divergent genic adjustment through disruptive selection or of divergence arising from the prevention of gene flow by geographic separation or other non-genetical means.

In an inbreeding species bars to crossing of the kind we have been discussing would not be expected to develop, since the inbreeding device is itself a bar to all crossing. Nor would we expect to see the rise of hybrid incapacity since in so far as the inbreeding was effective in producing homozygosis there would be no variability on which disruptive selection could act to produce divergence. It is therefore not surprising to find that in genera such as *Tritium* and *Avena*, which are characteristically inbreeding, species crosses are easy to make and the hybrids are both vigorous and fertile except in so far as differences in level of polyploidy between the parents lead to mechanical disturbances of chromosome behaviour at meiosis and so produce sterility.

In outbreeding species on the other hand, both inbreeding depression and hybrid incapacity are to be found side by side, as in *Drosophila spp.* where on the one hand inbred lines show loss of vigour and fertility and on the other crosses between different populations can lead to breakdown of balance and fitness in subsequent generations as we have already seen. We are led therefore to the notion of a hybridity optimum (Darlington and Mather, 1949), individuals that are too homozygous as a result of inbreeding or too hybrid as a result of over-wide crossing being equally at a disadvantage by comparison with those whose level of hybridity falls within the optimal range. At the same time we see a dual restriction on mating; inbreeding being avoided by the various devices that we looked at in Chapter 6, and over-wide crosses being avoided by the bars to crossing with which we are now concerned. The hybridity optimum is thus matched, as indeed it must be, by a corresponding control of mating.

As we have seen, devices which reduce or prevent inbreeding may be of many kinds. So too, may the bars to crossing by which over-wide hybridization is reduced or, in speciation, prevented, and the very mechanisms which in one form can be seen at work as devices preventing inbreeding can in other forms be seen as bars to crossing. In Thoday and Gibson's experiments crossing between the selected H and L lines was avoided by discriminative mating among the flies – a type of behaviour which obviously could also serve as a means of avoiding inbreeding. In an experiment with maize, quoted by Thoday (1972), Paterniani selected for the reduction of the frequency of hybrid grains between two varieties which originally flowered at much the same time, and ended with the two varieties differing by a week in their mean times of flowering. Correspondingly, differences in flowering, not between varieties but between the male and female inflorescences of a plant, can regulate the risk of inbreeding by self-pollination. Again, as we saw in Chapter 6, the style of an angiosperm flower can, through the phenomenon of incompatibility, exercise the capacity to sieve out genetically too like pollen and so prevent self-pollination. It can similarly discriminate against the genetically too unlike pollen from over-wide crosses as we observe with inter-specific pollinations in *Petunia, Streptocarpus, Antirrhinum* (Darlington and Mather, 1949) and other genera. The style in Angiosperms affords the means, and doubtless evolved in the first place as affording the means, of regulating the mating system in favour of not too close and not too wide a relationship, just as mating discrimination, though less well-investigated and less well-understood, would seem capable of doing in bisexual animals.

9   Individuals and Populations

Units of variation and selection

Our discussion has centred on the genes and genic combination to be found in populations and on the effects of various factors – inbreeding, linkage, mutation, migration, drift and above all selection – on the frequencies, distribution and fates of these genes and genic combinations. At the same time, as has been pointed out, the significance of genes and genic combinations lies in the phenotypes in the production of which they act and interact with one another and with external agencies, and they are affected by selection only through the medium of these phenotypes. In this sense, therefore, we start by treating the individual phenotype as the unit of selective action: a better adapted and fitter individual will contribute more to posterity than a less well-adapted and less fit individual which coexists alongside it. Commonly this simple picture will suffice. An individual is, for example, on its own in standing up to the rigours of the physical environment, cold, heat, drought, flood or whatever it may be (except to the extent that parental care is exercised and that in some species, notably man, individuals combine together to avoid or ameliorate these hazards). The individual is then the unit of selection. The same will generally be true in respect of other things like securing a suitable habitat, acquiring the necessary food and so on, where competition among individuals comes into the picture. Even where parents provide in some way for their offspring, there will be selection among individual parents in respect of the advantage of that provision and later when they are no longer being provided for. Selection is still a matter of individuals, and from this point of view the population is seen as no more than an assembly of individuals, acting independently of one another, modified only by an overlap between the generations which broadly parallels their genetic relationship.

Within a plant species showing regular self-pollination, and hence consisting of virtually complete homozygotes, each individual is capable, independently of its fellows, of producing offspring genetically like itself and as adequate and fit as itself. Each individual is thus self-sufficient not merely selectively but also genetically. In an outbreeding species, however, homozygotes are genetically in-adequate. A measure of heterozygosity is required for fitness and crossing must take place among the members of the population to maintain the appropriate level of heterozygosity. Thus the genetical well-being of an individual depends on it having had not a single line of ancestors but a network of them, and the genetical well-being of its descendents depends on it itself being one member of a population together with others with which it can cross. So in an outbreeding species the individual is not an adequate unit genetically: the group or population is a unit also, in that it is necessary to carry the requirement of genes and genic variation. Co-operation among individuals, in the form of crossing, is necessary to maintain the production of genetically adequate offspring, and where outbreeding is obligatory, as with bisexuality or incompatibility, co-operation in the form of crossing is necessary to maintain the production of any individuals at all. In other words, however it may be in respect of other characters and function, in respect of reproduction the individual is incomplete.

Where, as in some species of higher animals, a male and female pair for life they jointly constitute a unit of reproduction, and co-operation within the new unit may grow, as polymorphisms may do, to embrace a wide range of functions including defence of the common territory, acquisition of food, care of the young and so on. The pair is then becoming a selective unit not only in respect of reproduction and genetical transmission but in a wider sense too, for the two mates' selective fates are linked together; but the mated pair has still not completely ousted the individual as the selective unit, since for that part of the life-cycle before pairing takes place the individual's fate must be independent of that of its prospective mate.

Where the sexes are separate but mating is promiscuous (and equally in self-incompatible plants) co-operation between the members of a pair is essential for effective reproduction and their genetical fates are linked together in respect of any offspring they may jointly leave; but they have not become a unit of selection since each may mate with other, possibly many other, members of the opposite kind. Nor, since the relation between them is impermanent, is there likely to be much division of labour between them in respect of other functions. In such a case the individual is still to be taken as the effective unit of selection.

Co-operation between individuals in reproduction in obligatory outbreeding species, depends on their differences; but individuals which are alike may be of help to another by combining in an impermanent, almost casual, but nevertheless important way for some purposes. Thus *Drosophila* larvae compete with one another for any limited resources in a culture and the more larvae there are the more severe the competition. Yet larvae in cultures which are too under-populated are less successful, because below a certain density the larvae are not capable of successfully coping with the growth of the spoilage organisms which if not adequately checked lead to deterioration of culture conditions. Kearsey (1965) reports several experiments which illustrate this point, the results of one of which are illustrated in Fig. 25. The greatest percentage of eggs developing into adults was obtained when five – seven eggs were placed in each culture, a smaller percentage being recovered as adults when fewer eggs (1 – 3) or more eggs (9 – 11) were used to set up the culture. A similar need for a number of eggs to secure successful development has been reported for blowflies by Cragg (1955), and other examples could be quoted. In all cases, however, the co-operative effect appears to be unspecific and impermanent.

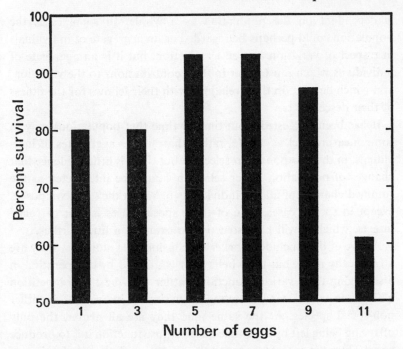

Fig. 25 Survival to the adult state of larvae in *Drosophila melanogaster* (Kearsey, 1965). The percentage survival is shown against the density of the larval population which was varied by placing different numbers of eggs in standard sized cultured vials. Maximum survival was secured at intermediate densities. When there were more larvae present in the vial they evidently competed with one another, but up to the optimal density they co-operated or helped one another.

It is thus unlikely not to leave the individual as the effective unit of selection.

There is thus something of a paradox in the situation. Despite co-operation in reproduction or other function, in the vast majority of cases the individual remains substantially the selective unit. At the same time, in outbreeding species the population is the unit for genetical variation. So in testing the fitness of genic combinations natural selection acts on individuals, each of which is effectively independent of its fellow; but the production of the genic combinations which are tested cannot be achieved by single individuals and is dependent on the population as a whole. In some cases the population could perhaps be regarded as an aggregate of individuals in respect of variation as well as selection, but it is an aggregate of individuals which must differ in their contributions to the variation, and which depend on their relations with their fellows for the fitness of their descendents.

It has been suggested from time to time that populations may in some measures act as whole, rather than just as aggregates of individuals, in their responses to selection but there is little evidence that changes of population under selection cannot be interpreted as the summed changes of all the individuals of which they are composed, except in special cases. One of these special cases is that of social insects, which it will be instructive to examine a little further.

A hive of bees includes workers in considerable numbers. They are all basically alike, but their behaviour is marked by co-operation in discharging their various functions rather than by the competition which one would expect among such similar individuals in other non-social species. At the same time they are all neuter, the only offspring being left by the single queen whose function it is to produce eggs. Thus the maintenance of the colony and the production of daughter colonies depends on the reproductive capacity of the queen. Hence competition among the workers could in no way increase their fitnesses as individuals, which is zero in any case. Co-operation among them in caring for the colony, supplying the queen, and nurturing the young on the other hand, has the advantage of raising the reproductive capacity of the colony which resides in the single queen, who however is dependent on the workers in discharging this function. The colony or hive is thus the unit of reproduction and of effective selection. The drones are of course in different case from the workers. Their contribution to the well-being of the colony is to facilitate the queen's reproductive capacity by fertilizing her on the mating flight. In doing this they are in competition with one another and, in view of the essentially co-operative nature of the colony's other activities, it is perhaps not surprising that after fertilization has been accomplished the competitive drones vanish from the scene.

As we have earlier seen, the polymorphism of the colony, and with it, as we now see, its functioning as a unit of selection, depends on the common genotype. This genotype must thus be under selection, through competition with other colonies, for the greater control of intra-hive competition and the more effective securing of co-operation among the hive's own constituent individuals. In this way a colony of bees, and similarly other social hymenopterans, is less like a population of a non-social species, among whose constituent individuals the relationship is basically one of competition, than it is like a differentiated soma.

In a differentiated soma, whether plant or animal, the cells, tissues and organs discharge different functions in relation to nutrition, locomotion, protection or whatever it may be. Among these cells, tissues and organs certain of them, like the queen in a colony of social insects, are responsible for reproduction. Of these tissues capable of reproduction some, like plant meristems or haemopoeic tissue in vertebrates, are concerned with extension and maintenance of the soma while still others are concerned with its overall reproduction by sexual means – aspects which are not separable, or at any rate, not separated in the functioning of the queen bee. Such cells in the soma must continue in divisions in a regulated way to discharge their function, but other cells in other tissues must equally, as part of their differentiation, give up further division. Now if a cell which should not divide continues to do so or resumes doing so, or if a cell in a dividing tissue escapes from the regulation of its divisions, the cell itself will have an advantage over its fellows and will reproduce competitively at their expense. At the same time it will distort the balance of the soma as a whole and with it the essential co-operative relation of the different cell and tissues upon which the overall well-being of the soma depends. And since the soma reproduces as a whole, the fitness of the whole system will be reduced, possibly to zero. The cell has gained a competitive advantage at its own level but only at the expense of disadvantage at the higher level of the whole soma. Thus success at the higher level of the soma in competition with other somata depends on the control of competition and the furtherance of co-operation at the lower level of cell and tissue. The common genotype of the soma, which is transmitted to offspring by the reproductive organs on behalf of the soma as a whole, will thus be selected to control the competition and further the necessary co-operation at the lower internal level, just as in the insect colony.

Human populations and societies

In both differentiated soma and insect society we can see the key position played by the genotype in controlling competition, while furthering co-operation, to secure the integration which gives soma or colony its chance of success in competition with other like somata or societies; and at the same time we can see how natural selection at this higher level adjusts the genotype to this end. In human societies integration is achieved in a different way. Man's capacity to communicate, initially through the spoken word but later strengthened by the development of means of recording communications first in writing and now by other means, has given him the ability to transmit information and ideas which result in persistent changes in the activities, operation and relationships of individuals within the society. As a result he has developed a form of extra-somatic evolution, by which he can change the environment to suit his needs or, through the use of machines, achieve for example flight, rapid locomotion, or the ability to live in water without need for the wings, special limbs or gills which other species employ for their purposes. This extra-somatic evolution depends on ideas and modes of thought just as somatic evolution depends on genes, and the ideas have the properties of transmission and variation (or mutation if we prefer it) and are subject to selection, just as genes are (see Mather 1964). The transmission of ideas is not however restricted to the parent-offspring relation. They can spread therefore much more rapidly than can genes (indeed the spread of ideas has more affinity with infection than heredity), and because they can spread more rapidly they permit a much more rapid evolution, with new ideas building up on top of one another in a way not limited by the time required for one generation to succeed another.

This same process of social transmission, through the medium of education (using this term in its broadest sense to cover not only formal education but the less formal transmission from parents, family and other members of the community which starts the moment a child is born) is the major means by which the individual is integrated into society and by which society itself evolves, again at a speed which biolgical transmission alone would not permit. Both the integration of the individual and the evolution of society require the regulation of competition and the fostering of appropriate co-operation in ways which become increasingly intricate as society and the sub-societies it contains becomes more complex. These complexities are examined in more detail elsewhere (Mather, *l.c.*) and need not detain us now. Our present point is that in human society social control and social evolution have overlaid the genetic control and genetic evolution that we see in other species and in so doing have opened up new dimensions of complexity and speed of change.

Behind the social transmission and evolution, however, the genetic system is still exercising its influence. Genetic differences exist in human populations as in other species of plants and animals. They are as ubiquitous and as subject to the action of selection. This is relatively readily demonstrable and is relatively readily accepted in respect of physical and physiological characters. It is less readily demonstrable and is much less readily accepted for the mental characteristics of ability and behaviour, for here the genetical basis is observed by the consequences of the social transmission built upon it. The evidence is nevertheless clear: ability as measured by I.Q. not only varies within human populations, but more than half of the variation is traceable to genetic differences. We have as yet no real understanding of how selection is affecting these genetic differences in ability, yet it is on the ability of individuals to communicate, of some to innovate and all to learn, that the new social evolution depends; and this ability, determined by the genotype, must have been built up by natural selection in the past. As we have seen, given that these genetic requirements are met, future social evolution need not depend on, nor its scope and speed be limited by, genetical selection and adjustment; but the maintenance of a minimum genetic level adequate to sustain social transmission is essential (see Mather, *l.c.*).

The interplay of the social and the genetic has been both fundamental and complex in the evolution of man and his society (Darlington, 1969). The social is built on the genetic and in its turn reacts back on the genetic. We have already seen one example of this in the use of social measures and social action to secure the outbreeding on which depends the biological fitness of the individuals, and hence in one of its essential aspects the well-being of the society (p. 111). In the same way social action can and indeed, whether premeditatedly or not, inevitably will alter the forces of selection impinging on the individual and hence result in change in the genetical composition of the population. In its earliest days, human groupings must have been small and based on the family, with the social transmission of education following much the same broad path as the reproductive transmission of genes. In such a case the two will largely go hand in hand, the social acting as an extension and fortification of the genetic, in binding the group together, determining its fitness and securing the survival of its social traditions as well as of its genes in competition with similar groupings. Indeed this situation will persist even in larger groupings so long as, and to the extent that, the social transmissions of education stays chiefly within the family grouping and hence with the genes. There is then no conflict in selection between the social and the genetic. But when, as is increasingly the case as society evolves, social transmission through education in its various forms becomes the task of a special group or groups within the society, the intellectual reproduction of the educator is no longer confined to his own biological offspring or those of his own family grouping, and the biological offspring of an individual to a corresponding extent cease to be his intellectual or social offspring. The two types of reproduction, biological and social, no longer march together. The selective consequences of this weakening of the correlation between the genetic and the social have hardly yet begun to be discussed, still less understood, yet they are basic, whether for good or for ill, to the future of human society.

But of one thing we may be confident. The same principles of transmission, variation and selection which we have seen working in terms of genes at the various levels of genetic integration, will still continue to appear as society develops though with an interplay made even more complex by the addition of the social elements and aspects to the genetic.

References

R. W. ALLARD, G. R. BARBEL, M. T. CLEGG and A. L. KAHLER (1972), 'Evidence for co-adaptation in *Avena barbata*', *Proc. Nat. Acad. Sci.*, Wash., **69**, 3043 – 3048.

R. W. ALLARD and D. R. MARSHALL (1969), 'The genetics of electrophoretic variants in *Avena*', *J. Hered.*, **60**, 17 – 19.

J. A. ALLEN (1972), 'Evidence for stabilizing and apostatic selection by wild blackbirds'. *Nature*, **237**, 348 – 349.

A. C. ALLISON (1955), 'Aspects of polymorphism in man', *Cold Spring Harb. Symp. Quant. Biol.*, **20**, 239 – 252.

E. ANDERSON (1949), *Introgressive Hybridization*, Wiley, New York.

F. J. AYALA, J. R. POWELL and M. L. TRACEY (1972), 'Enzyme variability in the *Drosophila willistoni* group. V. Genic variation in natural populations of *Drosophila equinonalis*', *Genet. Res.*, **20**, 19 – 42.

U. W. AYONDADU and H. REES (1971), 'The effects of B chromosomes on the nuclear phenotype in root meristems of maize', *Heredity*, **27**, 365 – 383.

B. W. BARNES (1968), 'Stabilizing selection in *Drosophila melanogaster*', *Heredity*, **23**, 433 – 442.

R. J. BERRY (1972), 'Genetical effects of radiation on populations', *Atomic Energy Rev.*, **10**, 67 – 100.

R. J. BERRY and J. H. CROTHERS (1968), 'Stabilizing selection in the dogwhelk (*Nucella lapillus*)', *J. Zool.*, **155**, 5 – 17.

W. F. BODMER and P. A. PARSONS (1962), 'Linkage and recombination in evolution', *Adv. Genet.*, **11**, 1 – 100.

A. D. BRADSHAW (1971), 'Plant evolution in extreme environments', *Ecological Genetics and Evolution*, ed. R. Creed, Blackwell, Oxford, 20 – 50.

E. L. BREESE (1959), 'Selection for different degrees of outbreeding in *Nicotiana rustica*'. *Ann. Bot.*, **23**, 331 – 344.

E. L. BREESE and K. MATHER (1957), 'The organization of polygenic activity within a chromosome in *Drosophila* I. Hair characters', *Heredity*, **11**, 373 – 395.

E. L. BREESE and K. MATHER (1960), 'The organization of polygenic activity within a chromosome in *Drosophila* II. Viability', *Heredity*, **14**, 375 – 399.

D. BRIGGS and S. M. WALTERS (1969), *Plant Variation and Evolution*, World University Library, London.

H. C. BUMPUS (1899), 'The elimination of the unfit as illustrated by the introduced sparrow'. *Biol. Lect. Woods Hole*, **1898**, 209 – 226.

D. H. CARR (1972), 'Chromosomal anomalies in human foetuses', *Res. in Reproduction*, **4**, (2), 3 – 4.

C. O. CARTER (1961), 'The inheritance of pyloric stenosis', *Br. Med. Bull.*, **17**, 251 – 253.

C. E. CATEN (1972), 'Vegetative incompatibility and cytoplasmic infection in fungi', *J. Gen. Microbiol.*, **72**, 221 – 229.

C. A. CLARKE and P. M. SHEPPARD (1960), 'The evolution of mimicry in the butterfly *Papilio dardanus*', *Heredity*, **14**, 163 – 173.

C. A. CLARKE and P. M. SHEPPARD (1962), 'Disruptive selection and its effect on a metrical character in the butterfly *Papilio dardanus*', *Evolution* **16**, 214 – 226.

G. CLAYTON and A. ROBERTSON (1955), 'Mutation and quantitative variation', *Am. Nat.*, **89**, 151 – 158.

M. T. CLEGG and R. W. ALLARD (1972), 'Patterns of genetic differentiation in the slender wild oat species *Avena barbata*', *Proc. Nat. Acad. Sci. Wash.*, **69**, 1820 – 1824.

L. M. COOK (1971), *Coefficients of Natural Selection*, Hutchinson, London.

P. COOKE and K. MATHER (1962), 'Estimating the components of continuous variation II. Genetical', *Heredity*, **17**, 211 – 236.

K. W. COOPER (1937), 'Reproductive behaviour and haploid parthoengenesis in the grass mite, *Pediculopsis graminum*', *Proc. Nat. Acad. Sci. Wash.*, **23**, 41 – 44.

W. M. COURT BROWN (1967), *Human Population Cytogenetics*, North Holland Pub. Co., Amsterdam.

J. B. CRAGG (1955), 'The natural history of sheep blowflies in Britain', *Ann. appl. Biol.*, **42**, 197 – 207.

R. CREED (1971), *Ecological Genetics and Evolution*, Blackwell, Oxford.

J. H. CROFT and G. SIMCHEN (1965), 'Natural variation among monokaryons of *Collybia velutipes*', *Am. Nat.*, **94**, 451 – 462.

J. F. CROW and M. KIMURA (1970), *An Introduction to Population Genetics Theory*, Harper and Row, New York.

C. D. DARLINGTON (1937), *Recent Advances in Cytology* (2nd. edn.), Churchill, London.

C. D. DARLINGTON (1939), *The Evolution of Genetic Systems*, Cambridge Univ. Press.

C. D. DARLINGTON (1963), *Chromosome Botany and the Origin of Cultivated Plants* (2nd edn.), Allen and Unwin, London.

C. D. DARLINGTON (1969), *The Evolution of Man and Society*, Allen and Unwin, London.

C. D. DARLINGTON and K. MATHER (1949), *The Elements of Genetics*, Allen and Unwin, London.

R. W. DAVIES (1971), 'The genetic relationship of two quantitative characters in *Drosophila melanogaster*. II. Location of the effects', *Genetics*, **69**, 363 – 375.

G. DEAN (1969), 'The porphyrias', *Br. Med. Bull.*, **25**, 48 – 51.

J. A. DETLEFSEN and E. ROBERTS (1921), 'Studies in crossing-over. I. The effect of selection on cross-over values'. *J. exp. Zool.*, **32**, 333 – 354.

T. H. DOBZHANSKY (1948), 'Genetics of natural populations. XVIII. Experiments on chromosomes of *Drosophila pseudoobscura* from different geographic regions', *Genetics*, **33**, 588 – 602.

T. H. DOBZHANSKY (1950), 'Genetics of natural populations. XIX. Origin of heterosis through natural selection in populations of *Drosophila pseudoobscura*', *Genetics*, **35**, 288 – 302.

T. H. DOBZHANSKY (1951), *Genetics and the Origin of Species* (3rd edn.), Columbia University Press, New York.

T. H. DOBZHANSKY (1971), 'Evolutionary oscillations in *Drosophila pseudoobscura*' *Ecological Genetics and Evolution*, ed. R. Creed, Blackwell, Oxford, 109 – 133.

T. H. DOBZHANSKY and S. WRIGHT (1947), 'Genetics of natural populations. XV. Rate of diffusion of a mutant gene through a population of *Drosophila pseudoobscura*' *Genetics*, **32**, 303 – 324.

A. DURRANT (1962), 'The environmental induction of heritable change in *Linum*', *Heredity*, **17**, 27 – 61.

L. EHRMAN and C. PETIT (1968), 'Genotype frequency and mating success in the *willistoni* species group of *Drosophila*', *Evolution*, **23**, 59 – 64.

D. S. FALCONER (1960), *'Introduction to Quantitative Genetics'*, Oliver and Boyd, Edinburgh.

D. S. FALCONER (1971), 'Improvement of litter size in a strain of mice at selection limit', *Genet. Res.*, **17**, 215 – 235.

G. FANKHAUSER (1941), 'The frequency of polyploidy, etc., in the newt', *Proc. Nat. Acad. Sci. Wash.*, **27**, 507 – 512.

R. A. FISHER (1930), *The Genetical Theory of Natural Selection*, Clarendon Press, Oxford.

R. A. FISHER (1949), *The Theory of Inbreeding*, Oliver and Boyd, Edinburgh.

R. A. FISHER and E. B. FORD (1947), 'The spread of a gene in natural conditions in a colony of the moth *Panaxia dominula* L.', *Heredity*, **1**, 143 – 174.

E. B. FORD (1945), 'Polymorphism', *Biol. Rev.*, **20**, 73 – 88.

E. B. FORD (1971), *Ecological Genetics* (3rd edn.), Chapman and Hall, London.

H. D. FORD and E. B. FORD (1930), 'Fluctuations in numbers and its influence on variation in *Melitaea aurinia*', *Trans. Roy. Ent. Soc. Lond.*, **78**, 345 – 351.

J. S. GALE and A. E. ARTHUR (1972), 'Variation in wild populations of *Papaver dubium*. IV. A survey of variation', *Heredity*, **28**, 91 – 100.

H. GRÜNEBERG (1938), 'An analysis of the "pleiotropic" effects of a new lethal mutation in the rat (*Mus norvegicus*)', *Proc. Roy. Soc.*, *B*, **125**, 123 – 143.

J. B. S. HALDANE (1957), 'The cost of natural selection', *J. Genet.*, **55**, 511 – 524.

D. J. HARBERD (1961), 'Observations on population structure and longevity of *Festuca rubra* L'. *New Phytol.*, **60**, 184 – 206.

H. HARRIS (1969), 'Genes and isozymes', *Proc. Roy. Soc.*, *B.*, **174**, 1 – 31.

H. HARRIS (1971), 'Polymorphism and protein evolution. The neutral mutation – random drift hypothesis', *J. Medical Genetics*, **8**, 444 – 452.

G. M. L. HASKELL (1949), 'Variation in the number of stamens in the common chickweed', *J. Genet.*, **49**, 291 – 301.

B. I. HAYMAN and K. MATHER (1953), 'The progress of inbreeding when homozygotes are at a disadvantage'. *Heredity*, **7**, 165 – 183.

J. HILL (1965), 'Environmental induction of heritable changes in *Nicotiana rustica*', *Nature*, **207**, 732 – 734.

S. L. HUANG, M. SINGH and K. KOJIMA (1971), 'A study of frequency-dependent selection observed in the esterase-6 locus of *Drosophila melanogaster* using a conditioned media method', *Genetics*, **68**, 97 – 104.

F. JACOB and J. MONOD (1963), 'Genetic repression, allosteric inhibition and cellular differentiation', *Cytodifferentiation and Macromolecular Synthesis*, 21*st Growth Symposium*, Academic Press, New York.

P. A. JACOBS, A. FRACKIEWICZ and P. LAW (1972), 'Incidence and mutation rates of structural rearrangements of the autosomes in man', *Ann. Human Genet.*, **35**, 301 – 314.

J. L. JINKS (1964), *Extrachromosomal Inheritance*, Prentice-Hall, Englewood Cliffs, N.J.

J. L. JINKS, C. E. CATEN, G. SIMCHEN and J. H. CROFT (1966),'Heterokaryon incompatibility and variation in wild populations of *Aspergillus nidulans*', *Heredity*, **21**, 227 – 239.

D. A. JONES and D. A. WILKINS (1971), *Variation and Adaptation in Plant Species*, Heinemann, London.

R. N. JONES and H. REES (1969), 'An anomalous variation due to B chromosomes in rye', *Heredity*, **24**, 265 – 271.

W. JOHANNSEN (1909), *Elemente der exakten Erblechkeitslehre*, Fischer, Jena.

M. N. KARN and L. S. PENROSE (1952), 'Birth weight and gestation time in relation to maternal age, parity and infant survival', *Ann. Eugen.*, **16**, 147 – 164.

M. J. KEARSEY (1965), 'Co-operation among larvae of a wild type strain of *Drosophila melanogaster*', *Heredity*, **20**, 309 – 312.

M. J. KEARSEY and B. W. BARNES (1970), 'Variation for metrical characters in *Drosophila* populations. II. Natural selection', *Heredity*, **25**, 11 – 21.

N

M. J. KEARSEY and K. KOJIMA (1967), 'The genetic architecture of body weight and egg hatchability in *Drosophila melanogaster*', *Genetics*, **56**, 23 – 37.

H. B. D. KETTLEWELL (1956), 'Further selection experiments on industrial melanism in the Lepidoptera', *Heredity*, **10**, 287 – 301.

M. G. KIDWELL (1972), 'Genetic change of recombination value in *Drosophila melanogaster* I and II', *Genetics*, **70**, 419 – 432 and 433 – 443.

M. J. LAWRENCE (1972), 'Variation in wild populations of *Papaver dubium* III', *Heredity*, **28**, 71 – 90.

I. M. LERNER (1954), *Genetic Homeostasis*, Oliver and Boyd, Edinburgh.

D. LEWIS (1954), 'Comparative incompatibility in angiosperms and fungi', *Adv. Genet.*, **6**, 235 – 287.

K. R. LEWIS and B. JOHN (1963), *Chromosome Marker*, Churchill, London.

C. C. LI (1967), 'Genetic equilibrium under selection', *Biometrics*, **23**, 397 – 484.

K. MATHER (1938), 'Crossing-over', *Biol. Rev.*, **13**, 252 – 292.

K. MATHER (1940), 'The determination of position in crossing-over, III. The evidence of metaphase chiasmata', *J. Genet.*, **39**, 205 – 223.

K. MATHER (1941), 'Variation and selection of polygenic characters', *J. Genet.*, **41**, 159 – 193.

K. MATHER (1943), 'Polygenic inheritance and natural selection', *Biol. Rev.*, **18**, 32 – 64.

K. MATHER (1948), 'Nucleus and cytoplasm in differentiation', *Symp. Soc. Exp. Biol.*, **2**, 196 – 216.

K. MATHER (1949), *Biometrical Genetics* (1st edn.) Methuen, London.

K. MATHER (1950), 'The genetical architecture of heterostyly in *Primula sinensis*', *Evolution*, **4**, 340 – 352.

K. MATHER (1951), *The Measurement of Linkage in Heredity* (2nd edn.), Methuen, London.

K. MATHER (1953), 'The genetical structure of populations', *Symp. Soc. Exp. Biol.*, **7**, 66 – 95.

K. MATHER (1960), 'Evolution in polygenic systems', *Evoluzione e Genetica*, 131 – 152, Academia Nazionale dei Lincei, Rome.

K. MATHER (1961), 'Genetics', *Contemporary Botanical Thought*, ed. A.M. Macleod and L. S. Cobley, Oliver and Boyd, Edinburgh, 47 – 94.

K. MATHER (1964), *Human Diversity*, Oliver and Boyd, Edinburgh.

K. MATHER (1969), 'Selection through competition', *Heredity*, **24**, 529 – 540.

K. MATHER and P. M. J. EDWARDES (1943), 'Specific differences in *Petunia* III Flower colour and genetic isolation', *J. Genet.*, **45**, 243 – 260.

K. MATHER and B. J. HARRISON (1949), 'The manifold effect of selection', *Heredity*, **3**, 1 – 52, 131 – 162.

K. MATHER and J. L. JINKS (1958), 'Cytoplasm in sexual reproduction', *Nature*, **182**, 1188 – 1190.

K. MATHER and J. L. JINKS (1971), *Biometrical Genetics* (2nd. edn.) Chapman and Hall, London.

K. MATHER and A. VINES (1951), 'Species crosses in *Antirrhinum* II. Cleistogamy in the derivatives of *A. majus* and *A. glutinosum*', *Heredity* **5**, 195 – 214.

K. MATHER and L. G. WIGAN (1942), 'The selection of invisible mutations', *Proc. Roy. Soc. B.*, **131**, 50 – 64.

E. MAYR (1954), *Animal Species and Evolution*, Harvard Univ. Press, Cambridge, Mass.

A. MCGILL and K. MATHER (1971), 'Competition in *Drosophila* I. A case of stabilizing selection'. *Heredity*, **27**, 473 – 478.

MEDICAL RESEARCH COUNCIL (1956), *The Hazards to Man of Nuclear and Allied Radiations*. H.M.S.O. (Cmd. 9780), London.

E. T. MØRCH (1941), *Chondrodystrophic Dwarfs in Denmark*, Eynar Munksgaard, Copenhagen.

H. J. MULLER (1932), 'Further studies on the nature and causes of gene mutations', *Proc. 6th Int. Cong. Genetics*, **1**, 213 – 255.

H. J. MULLER (1950), 'Our load of mutations', *Am. J. Human Genet.*, **2**, 111 – 176.

A. MÜNTZING (1963), 'The effects of accessory chromosomes in diploid and tetraploid rye', *Hereditas*, **49**, 371 – 426.

H. R. NEVANLINNA (1972), 'The Finnish population structure. A genetic and genealogical study', *Hereditas*, **71**, 195 – 236.

H. NILSSON-EHLE (1912), *Kreuzunguntersuchungen an Hafer und Weizen*, Lund.

A. NYGREN (1966), 'Apomixis in the angiosperms, with special reference to *Calamagrostis* and *Poa*', *Reproductive Biology and Taxonomic Behaviour of Vascular Plants* (ed. J. G. Hawkes), Pergamon Press, London, 131 – 140.

R. ORNDUFF (1972), 'The breakdown of trimorphic incompatibility in *Oxalis* Section Corniculatae', *Evolution*, **26**, 52 – 65.

J. A. PATEMAN and B. T. O. LEE (1960), 'Segregation of polygenes in ordered tetrads', *Heredity*, **15**, 351 – 361.

E. PATERNIANI (1969), 'Selection for reproductive isolation between two populations of maize, *Zea mays* L.', *Evolution*, **23**, 534 – 547.

G. J. PAXMAN (1957), 'A study of spontaneous mutation in *Drosophila melanogaster*', *Genetica*, **29**, 39 – 57.

P. E. POLANI (1967), 'Occurrence and effect of human chromosome abnormalities', *Social and Genetic Influences on Life and Death, Eugenics Soc. Symposia*, **3**, 3 – 19.

J. R. POWELL (1971), 'Genetic polymorphism in a varied environment', *Science*, **174**, 1035 – 1036.

J. R. RAPER, G. S. KRONGELB and M. G. BAXTER (1958), 'The number and distribution of incompatibility factors in *Schizophyllum*', *Am. Nat.* **92**, 221 – 232.

H. REES (1955), 'Genotypic control of chromosome form and behaviour', *Bot. Rev.*, **27**, 288 – 318.

c. RICK (1945), 'A survey of cytogenetic causes of unfruitfulness in the tomato', *Genetics*, **30**, 347 – 362.

R. RILEY and V. CHAPMAN (1958), 'Genetic control of the cytologically diploid behaviour of hexaploid wheat', *Nature*, **182**, 713 – 715.

A. ROBERTSON (1955), 'Selection in animals: synthesis', *Cold Spring Harb. Symp. quant. Biol.*, **20**, 225 – 229.

R. K. SELANDER, S. Y. YANG, R. C. LEWONTIN and W. E. JOHNSON (1970), 'Genetic variability on the horseshoe crab (*Limulus polyphemus*), a phylogenetic "relic",' *Evolution*, **24**, 402 – 414.

P. M. SHEPPARD (1953), 'Polymorphism and population studies', *Symp. Soc. exp. Biol.*, **7**, 274 – 289.

G. SIMCHEN (1966), 'Monokaryotic variation and haploid selection in *Schizophyllum commune*', *Heredity*, **21**, 241 – 263.

G. SIMCHEN and J. STAMBERG (1969), 'Fine and coarse controls of genetic recombination', *Nature*, **222**, 329 – 332.

O. T. SOLBRIG (1972), 'Breeding system and genetic variation in *Leavenworthia*', *Evolution*, **26**, 155 – 160.

S. G. SPICKETT and J. M. THODAY (1966), 'Regular responses to selection. 3. Interaction between located polygenes', *Genet. Res.*, **7**, 96 – 121.

G. L. STEBBINS (1950), *Variation and Evolution in Plants*, Columbia Univ. Press, New York.

C. STERN (1960), *Human Genetics* (2nd edn.), Freeman, San Francisco.

A. C. STEVENSON (1961), 'Frequency of congenital and hereditary disease with special reference to mutations', *Br. Med. Bull.*, **17**, 254 – 259.

F. A. STREAMS and D. PIMENTAL (1961), 'Effects of immigration on the evolution of populations', *Am. Nat.*, **95**, 201 – 210.

J. M. THODAY (1961), 'Location of polygenes', *Nature*, **191**, 368 – 370.

J. M. THODAY (1972), 'Disruptive selection', *Proc. Roy. Soc. B.*, **182**, 109 – 143.

J. M. THODAY and T. B. BOAM (1961), 'Regular responses to selection. 1. Description of responses', *Genet. Res.*, **2**, 161 – 176.

M. VETUKHIV (1954), 'Integration of the genotype in local populations of three species of *Drosophila*', *Evolution*, **8**, 241 – 251.

C. H. WADDINGTON (1939), *An Introduction to Modern Genetics*, Allen and Unwin, London.

B. WALLACE (1970), '*Genetic Load*, Prentice-Hall, Englewood Cliffs, N.J.

B. WALLACE and M. VETUKHIV (1955), 'Adaptive organization of the gene pools of *Drosophila* populations', *Cold Spring Harb. Symp. quant. Biol.*, **20**, 303 – 309.

W. F. R. WELDON (1901), 'A first study of natural selection in *Clausilia laminata*', *Biometrika*, **1**, 109.

M. J. D. WHITE (1954), *Animal Cytology and Evolution* (2nd edn.) Cambridge Univ. Press.

D. WILKIE (1956), 'Incompatibility in bracken', *Heredity*, **10**, 247 – 256.

W. WILLIAMS and N. GILBERT (1960), 'Heterosis and the inheritance of yield in the tomato'. *Heredity*, **14**, 133 – 149.

M. WILLIAMSON (1959), 'Studies on the colour and genetics of the black slug', *Proc. Roy. Physical Soc. Edinb.* **27**, 87 – 93.

S. WRIGHT (1931), 'Evolution in mendelian populations', *Genetics*, **16**, 97 – 159.

S. WRIGHT (1940), 'The statistical consequences of mendelian heredity in relation to speciation', *The New Systematics*, 161 – 183, Oxford Univ. Press.

Index

achondroplasia, 40, 41
Allard, 56
Allen, 53
Allison, 2, 47
Allium, 124
amphidiploid, 8, 9
Anderson, 166
ants, 152
Antirrhinum, 117, 128, 171
aphids, 106, 153, 158
Arion, 107
Artemia, 105
apomixis, 102, 105, 106, 126
Arthur, 22
Aspergillus, 167
Avena, 16, 56, 107, 170
Ayala *et al.*, 56
Ayonoadu, 11

background (common, general)
    genotype, 110, 115, 116, 150,
    152, 153, 156, 159, 177, 178
bacteria, 1, 2
balance, 132–138, 142, 147
    internal ——, 132–134
    relational ——, 132–134, 165
barley, 107
Barnes, 92, 145
B chromosomes, 10–11, 20, 121
bees, 151, 153, 176

Berry, 21
Blood-groups, 20, 26
    ABO, 48
    MN, 26
    Rh, 48
blow-flies, 175
Boam, 85
Bodmer, 121
body weight, 145
Bonellia, 109, 150, 153, 158, 159
Bradshaw, 2, 163
breeding system, 106–118, 126, 127,
    128, 138, 142
    stratified ——, 117
Breese, 86, 107, 131, 144, 145, 147
Bumpus, 90

Campanula, 19
Carr, 9, 17, 18
Carter, 42
cat, 28, 29
Caten, 167
chaetae
    abdominal ——, 85, 95, 131, 140,
    141, 144, 145
    —— number, 144
    sternopleural ——, 85, 92, 93,
    131, 140, 145, 146, 154, 161,
    162, 168
Chapman, 9

191

chemical mutagens, 5, 41
chiasma frequency, 121–122
      position, 122–125
chromosome system, 126, 127, 142
Clarke, 20, 151
Clarkia, 11
Clayton, 81, 86, 95
Clegg, 56
cleistogamy, 107
co-adaptation, 134–136, 165, 166
colour-blindness, 28–29
competition *v.* selection,
      competitive
Cook, 53, 54, 56
Cooper, 31
co-operation, 173–178
correlated responses, 139–142
Court Brown, 12
Cragg, 175
Croft, 22
crossing, 63, 64, 72, 84, 103, 104,
      106, 118, 164, 165
Crow, 54

Darlington, 1, 11, 17, 19, 29, 102,
      105, 109, 111, 115, 120, 121,
      124, 125, 126, 134, 156, 158,
      170, 171, 180
Darwin, 2, 23, 113
Datura, 10
Davies, 86, 140
D.D.T., 1, 2
Dean, 84
Dendraster, 152
Detlefsen, 122
developmental plasticity, 158–159
diabetes, 42
differentiation, 152, 153, 177
dioecy, 108, 110
divergence, 161–163, 164, 165, 168
DNA, 14, 15

Dobzhansky, 1, 3, 19, 21, 82, 121,
      134, 136
dominance, 79–80, 144–147
drift, 82–84
Drosophila, 1, 6, 10, 15, 17, 19, 20,
      21, 53, 81, 119, 121, 125, 129,
      130, 131, 142, 145, 146, 162,
      167, 174
—— *equinonalis*, 56
—— *melanogaster*, 12, 85, 92, 93,
      122, 124, 140, 144, 154, 161,
      168, 175
—— *paulistorum*, 136
—— *persimilis*, 19
—— *pseudoobscura*, 3, 19, 82,
      134, 135, 136
Durrant, 5

ecological niches, *v.* environmental
      niches
Edwardes, 165
egg hatchability, 145
Ehrman, 53
environmental change, 97–100, 157
   abrupt ——, 99–100, 101, 143,
      166
   cyclical ——, 97–99, 100, 101,
      142, 157
   fluctuating ——, 97–99, 100, 101
   trend ——, 99–100, 101
environmental niches, 96, 149–150,
      161
environmental variation, 159–160
Epilobium, 6, 107
Escherichia, 16, 152
evolution, 3, 178
   extra-somatic ——, 178
   social ——, 179

Falconer, 28, 80, 86, 95, 96
Fankhauser, 17
fertility, 129–131, 140–142, 144

Festuca, 102
Fisher, 31, 83, 94, 96, 144, 145, 155
fitness, 2, 44, 57, 60, 92, 94, 131, 134, 143–144, 164, 173, 175, 177
flax, 5
foetuses, 9, 17
Ford, 20, 58, 83, 116
founder principle, 84
french beans, 30
Fritillaria, 123

Gale, 22
gene(s)
    classification of ——, 14–16
    combinations of ——, 87, 136, 138, 145, 165, 175
    —— complexes, 120
    —— conversion, 119
    —— flow, 161–163, 164, 165, 166, 167, 169
    lethal ——, 15, 21
    sex-linked ——, 27–29
    super ——, 120, 138, 153, 155, 157
genetic architecture, 143–147
genetic deaths, 59
genetic system, 125–128
genic interaction, 144–147
geographical separation, 165, 166, 167
Gibson, 167, 168, 171
Gilbert, 131
grasshoppers, 19, 123
Grüneberg, 16

haemophilia, 20
Haldane, 58
Harberd, 102
Hardy-Weinberg, 24–29
Harris, 1, 20, 56

Harrison, 85, 86, 140, 141, 143, 145, 147
Haskell, 30
Hayman, 35
heritability, 95, 96
heterosis, 129–131
heterostathmy, 108
heterostyly, 113–114, 116, 154, 156
heterozygotes
    advantage of ——, 33–35, 47, 48, 130
    fitness of ——, 45–47
    maximum proportion of ——, 35–37, 108
Hill, 5
homostyles, 116, 118, 155, 156
house flies, 1, 2
Huang, 53
human societies, 178–181
hybridity optimum, 170
Hymenoptera, 110, 153

ideas, 178
immigration, 42, 59
inbreeding, 29–37, 104, 106, 107, 110, 116, 127, 129–134, 138, 170
    —— depression, 121, 130, 170
    —— species, 131 *et seq.*, 170, 173
incest, 117
incompatibility, 51–53, 111–116, 150, 153, 155, 171, 173
    gametophytic ——, 111–112
    heterokaryon ——, 167
    sporophytic ——, 112–113
inertia, 138–143, 147
integration, 138–143
industrial melanism, 56
interchange, 12–13, 19, 120, 125
introgression, 166
inversion, 12–13, 19, 125, 134–135, 155

ionizing radiations, 5, 41
I.Q., 179
isolating mechanisms
  bars to crossing, 164–170
  hybrid incapacity, 165, 169–170
isolation, 165–168

Jacob, 16, 152
Jacobs *et al.*, 19
Jinks, 5, 17, 68, 80, 102, 167
Johannsen, 30
John, 17, 19, 121
Jones, D. A., 102
Jones, R. N., 11

Karn, 90, 91, 94
Kearsey, 92, 145, 174, 175
Kettlewell, 20
Kidwell, 122, 124
Kimura, 54
Kojima, 53, 145

Landsteiner, 26
Lawrence, C. W., 122
Lawrence, M. J., 22
Leavenworthia, 118
Lebistes, 109
Lee, 22
Lerner, 130
Levine, 26
Lewis, D., 51, 111, 114
Lewis, K. R., 17, 19, 121
Li, 54
life cycle, length of, 100, 101, 127,
  142
Limulus, 20
linkage, 74–75, 142
Linum, 115
load
  genetic ——, 54–60
  mutational ——, 54–55
  segregational ——, 55–56

locusts, 153
Lychnis, 110, 122
Lythrum, 113

maize, 108, 129, 171
malaria, 2, 47
man, 9, 10, 19, 20, 28, 40, 47, 48,
  109, 111, 117, 178–181
  birthweight and neo-natal
    mortality in ——, 90–92, 94
Mather, 17, 22, 23, 26, 29, 35, 43,
  48, 50, 57, 59, 68, 78, 80, 85,
  86, 90, 92, 93, 94, 102, 110,
  111, 115, 120, 122, 123, 124,
  125, 128, 131, 134, 140, 141,
  143, 144, 145, 147, 152, 156,
  158, 165, 170, 171, 178, 180
mating
  assortative ——, 31–33
  cousin ——, 31
  —— discrimination, 168, 171
  half-sib ——, 31, 35
  parent-offspring —— 30,
  random ——, 24, 25, 28, 36–37,
    64, 65
  sib ——, 30, 31, 35, 107, 116
Mayr, 84, 167
McGill, 92, 93, 94
Medical Research Council, 41
Melitaea, 58
meiotic drive, 119–120
Mendel, 30
metal toxicity, 2
mice, *t* alleles, 119–120
mimicry, 151
mongols, 12, 19
Monod, 16, 152
monoecy, 108
Mørch, 40, 41
Morris, 95
mosquitoes, 6

Muller, 14, 21
Müntzing, 11
mutation, 5, 19, 37–44, 59, 81, 86, 87
   back ——, 37

Nevanlinna, 84
newts, 17
Nicotiana, 5, 107
Nilsson-Ehle, 16
Nygren, 106

oats, *v. Avena*
Oenothera, 13, 19, 23, 51, 112, 120, 125, 127
optimal phenotype, 88–89, 92, 93, 96, 148–149, 164
Ornduff, 115
outbreeding (outcrossing), 107, 108, 116, 118, 127, 128, 138
   over-wide ——, 110, 116, 170
   —— species, 131 *et seq.*, 170, 173, 174, 175
Oxalis, 113, 115

Panaxia, 83–84
Papaver, 22
Papilio, 151
Parsons, 121
parthenogenesis, 102, 103, 105, 158
Pateman, 22
Paterniani, 171
Paxman, 81, 86
Pediculopsis, 31, 107, 110, 116
penicillin, 1
Penrose, 90, 91, 94
Petit, 53
Petunia, 171
phenylketonuria, 20
Phytophthora, 2
Pickford, 29
pigs, 134, 144

Pimental, 162, 163
plasmon, 5, 6
pleiotropy, 15–16, 140
Poa, 11
Polani, 9, 18
polygenic-systems, 16–17, 23, 69 *et seq.*, 86, 118, 132, 138, 142, 160
   —— combinations, 165, 166
   —— variability, 166
polymorphism, 20, 47, 55–56, 135, 149–157, 158, 162, 168, 173, 177
   quasi ——, 152, 153
   temporal (cyclical) ——, 159
polyploidy, 7 *et seq.*, 17, 22, 121, 126, 128
polysomy, 8 *et seq.*, 22
population size, 58, 59, 82–84
   —— of Finland, 84
porphyria, 84
potatoes, 1–2
poultry, 129
Powell, 160
Primula
   —— *sinensis* (Chinese primrose), 115, 156
   —— *vulgaris* (primrose) 113, 116, 155
protandry and protogyny, 107
Pteridium, 115
pyloric stenosis, 42

Ranunculus, 159
Raper *et al.*, 115
rat, 15
recombination, 13, 73–75, 84, 86, 87, 102, 103, 106, 121–125, 126, 128, 136, 137, 142, 155, 166
   —— index, 121
Rees, 9, 11, 121

reproduction
  asexual ——, 101–102, 104, 105,
    126–127
  biological and social ——, 180
  pseudo-sexual ——, 101–103,
    105, 126–127
  sexual ——, 101, 103–104, 105,
    126, 127
  sub-sexual ——, 103
  —— system, 127, 142
  units of ——, 173, 176
Rhoeo, 19
Rick, 18
Riley, 9
Roberts, 122
Robertson, 81, 86, 95, 144
rye, 121

Schizophyllum, 115
Sciara, 111, 117, 119, 120
Searle, 28
segregation, 63, 64, 72, 84, 102, 103,
    106, 118–120, 126, 128
Selander *et al.*, 20
selection, 2, 3, 5, 35, 38 *et seq.*,
    44–54, 65–67, 72, 122, 124, 126,
    139, 141, 142, 143–147, 153
  competitive ——, 43–44, 48–51,
    57, 58, 60, 92, 93, 177, 178
  density-dependent ——, 50–53,
    57, 58
  —— differential, 96
  directional ——, 92–96, 97, 100,
    101, 106, 125, 143–145
  disruptive ——, 96–97, 148–171
  —— experiments, 84–87, 94, 139
  frequency-dependent ——, 51–54
  fundamental theorem of
    natural ——, 94
  stabilizing ——, 88–92, 93, 97,
    100, 101, 106, 125, 143–145
  unconditional ——, 42, 57

units of ——, 172–176
self-fertilization (selfing), 30, 107,
    117, 170
sex
  -determination, 85, 109–110, 153,
    156
  -dimorphism, 149–151
  hetero- and homogametic ——,
    27
Sheppard, 20, 83, 151
sickle-cell anaemia, 2, 47–48, 56
Simchen, 22, 122
Singh, 53
snails, 90
social transmission in man, 178–181
Solbrig, 118
sparrows, 90
specialization of function, 153
speciation,
  allopatric ——, 167
  sympatric ——, 167, 168
Spickett, 85, 87, 146, 147
Stamberg, 122
Stebbins, 1, 17
Stellaria, 30
Stern, 1
Stethophyma, 123
Stevenson, 21
Streams, 162, 163
Streptocarpus, 171
Strongylocentrotus, 152
sugar beet, 129
switching mechanism, 110, 114,
    115–116, 150, 151, 152, 153,
    154, 156, 157

tetraploid, 8, 9, 17, 18, 105
Thoday, 85, 86, 87, 146, 147, 149,
    154, 155, 160, 161, 162, 163,
    167, 168, 171
tomatoes, 18, 107, 131
Tradescantia, 11

triploid, 8, 9, 17, 18, 104, 105
trisomic, 8, 10, 12, 18, 23
Triticum, 16, 107, 170
Tulipa, 105, 123

variability
  balance sheet of ——, 71, 77
  D and *H* ——, 80, 96
  fixed ——, 67, 80
  flow of ——, 142
  free ——, 61–64, 73, 76–78, 103
  heterozygotic potential ——,
    71–75, 76–78, 84, 103, 104
  homozygotic potential ——,
    71–75, 76–78, 84, 103, 104
  potential ——, 61–64
  utilized ——, 65–66
variation, 3, 4 *et seq.*, 89, 94, 100,
  126
  chromosomal ——, 6–14
  continuous ——, 16–17, 160
  cytoplasmic ——, 102

genic ——, 6, 14–17
heritable and non-heritable ——,
  4–6
numerical and structural ——, 6
Vetukhiv, 136
viability, 144
vigour, 129–131
Vines, 128

Wallace, 56, 57, 58, 136
Weldon, 90
Waddington, 152
wheat, *v. Triticum*
White, 1, 19, 109, 119
Wigan, 86
Wilkie, 115
Wilkins, 102
Williams, 131
Williamson, 107
Wright, 82

X-Y differences, 109